POTATO CHIP SCIENCE

A. Kurzweil & Son

Illustrations by Rob Walker

WORKMAN PUBLISHING · NEW YORK

For Robert Kurzweil (1911–1966)
Engineer, Inventor, Nurturer of Young Scientists

Library of Congress Cataloging-in-Publication Data is available.
ISBN 978-0-7611-4825-8

Illustrations: boy (front cover, bag) by Brian Biggs; chips (front
cover, bag), spot art (back cover), and interior art by Rob Walker;
author photo (back cover) by Constance Dussart
Book design by Rob Walker and Dave Kendrick
Art direction by Francesca Messina, Janet Vicario,
Janet Parker, Rae Ann Spitzenberger, and Rob Walker
Editorial coordinator: Alex Dunwoodie
Science advisor: Eric Kravitz

Workman books are available at special discounts when purchased in
bulk for premiums and sales promotions as well as for fund-raising or
educational use. Special editions or book excerpts can also be created
to specification. For details, contact the Special Sales Director at the
address below.

Workman Publishing Company
225 Varick Street
New York, NY 10014-4381
www.workman.com

Printed in the United States of America
First Printing September 2009
10 9 8 7 6 5 4 3 2

po·ta·to chip sci·ence \pə-tā'tō chip sī'əns\ *n*
the branch of science specifically devoted to the
composition, structure, reactions, and properties of
thin-sliced, deep-fried tubers, uncooked potatoes,
potato chip bags, lids, and tubes

CONTENTS

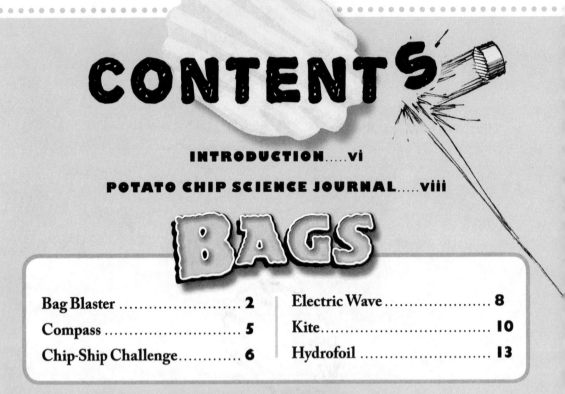

BAGS

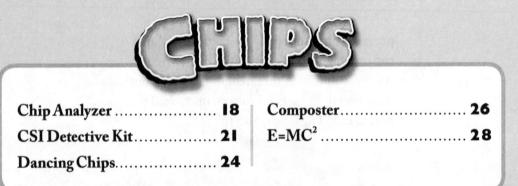

CHIPS

LIDS

INTRODUCTION

Yellow there! Welcome to Potato Chip Science, an experimental universe cooked up for folks who love science … and snacks … and snacking *on* science.

If that describes you, we're sure you'll devour this kit. Here's what it contains:

- *1 book that includes 29 snacktivities, 5 "spreads," 1 glossary, and 6 optical stickers.*
- *1 sound chip and 1 clock.*
- *1 biodegradable starch knife.*
- *1 propulsion pipe.*
- *Plus tons of other goodies and guidance to help you blast bags, burn chips, spin lids, and transform empty tubes into launchers. (We've even included the perfect ammo for your confetti can-non: eco-friendly packing "chips.")*

But hold on! Before you start studying the world's most miraculous munchie, here are a few more helpful morsels of information.

Morsel No. 1: How to Use the Kit

The arrangement of this book is every bit as nifty as its packaging. Chapters are divided among bags, chips, lids, spuds, and tubes. Why? Because these five branches of Potato Chip Science provide a *non*traditional way to study the five branches of traditional science. If there's a scientific term you're not familiar with, simply check the

Five Branches of Natural Science

(Traditional)	(Nontraditional)
Astronomy	Bags
Biology	Chips
Chemistry	Lids
Earth Science	Spuds
Physics	Tubes

glossary. (Geeks can relax: We've also included "old school" field and concept labels.)

The organization of the experiments within the five chapters is also a bit different. We've set up the directions using our straightforward *M-M-M-M* System of experimental inquiry. The first M stands for Materials: what you need. The second M stands for Method: what you have to do. The third M stands for Meaning: what's going on in the experiment. The final M stands for Munch on This: food for thought about the tests undertaken. (Oh, and in case you're wondering, the system is pronounced "*MMMM!*")

The *M-M-M-M* System

Materials = What you need.
Method = What you have to do.
Meaning = What's going on.
Munch on This = Food for thought.

Morsel No. 2: Recycling vs. Reuse

Before you begin to experiment, here are a few thoughts about the environmental implications of this science kit.

Potato Chip Science is *not* about recycling. It is about reuse. There's a big difference. Recycling requires old stuff to be treated at a processing plant, where it's turned into raw material that's then turned into *new* stuff.

Reuse skips all that. It simply takes old stuff and finds new things to do with it. What's so great about that? Plenty!

Potato chip packaging places a huge strain on our ecosystem. Every bag, tube, and lid used in a science experiment is one bag, tube, or lid that bypasses the landfill. So next time you're asked to take out the garbage, stop to inspect what's inside. For all you know, that trash pail contains an experiment waiting to crunch, pop, or fizzle, float, blast, or flip. (And if it doesn't, you can always reach for this kit.)

The one thing that *doesn't* come with this book is an adult to chip in with the trickier experiments. This brings us to the last morsel.

Morsel No. 3: Potato Chip Science Safety

You probably know the drill. But just in case you don't:

- Never eat research materials unless specifically instructed to do so.

- Whenever blasting, launching, or tossing objects, wear proper eye protection.

- Never blast, launch, or toss stuff at living creatures.

- Keep your hair tied back whenever conducting experiments.

- Potato chips and potato chip bags are HIGHLY flammable. Never let chip material of any kind get near fire (unless specifically instructed to do so and in the presence of an adult).

- And speaking of adults, here's one final note: When an experiment requires the assistance of a grown-up, we have included this warning:

AN ADULT MUST CHIP IN

While some of the projects in this book can be completed in just a few minutes, others take a bit longer to, well, digest. That's intentional. *Potato Chip Science* has been designed to provide hours and hours of spuddy study. In fact, all these experiments come with a single, simple challenge: We bet you can't try just one!

Chips away!

A. Kurzweil & Son

NOTABLE NOTES: A lab journal generally provides a record of how and why experiments are conducted, which methods and materials work, and which methods and materials do not. But it can do so much more! It can contain sketches and relevant newspaper articles, photographs, and charts. It can be a place to ask questions and to jot down ideas. And by keeping all the bits and pieces of your experimental curiosity in a single place, you will improve your chances of generating a breakthrough.

BAGS

→ INDUSTRY NAME: FLEXIBLE PACKAGING

→ SCIENTIFIC NAME: BIAXIALLY-ORIENTED POLYPROPYLENE (BOPP)

PROS
* FLEXIBILITY
* CLARITY
* KEEPS CHIPS FRESH
* PIZZAZZ
* LIGHTWEIGHT
* INEXPENSIVE
* WATERPROOF
* AIRTIGHT

CONS
* TOXIC GAS EMITTED DURING PRODUCTION
* SLOW DE-COMPOSITION
* ADDS <u>BIGTIME</u> TO LANDFILL!

DIMENSIONAL CHARACTERISTICS

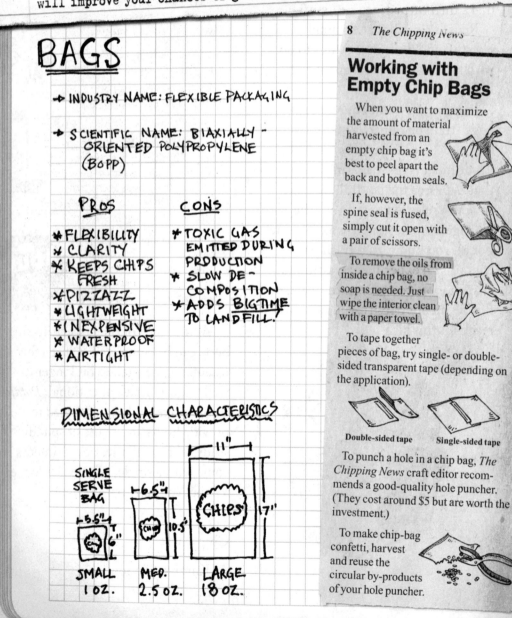

SINGLE SERVE BAG

├5.5"┤
6"
SMALL 1 OZ.

├6.5"┤
10.5"
MED. 2.5 OZ.

├ 11" ┤
CHIPS
17"
LARGE 18 OZ.

---

8   *The Chipping News*

Working with Empty Chip Bags

When you want to maximize the amount of material harvested from an empty chip bag it's best to peel apart the back and bottom seals.

If, however, the spine seal is fused, simply cut it open with a pair of scissors.

To remove the oils from inside a chip bag, no soap is needed. Just wipe the interior clean with a paper towel.

To tape together pieces of bag, try single- or double-sided transparent tape (depending on the application).

Double-sided tape   Single-sided tape

To punch a hole in a chip bag, *The Chipping News* craft editor recommends a good-quality hole puncher. (They cost around $5 but are worth the investment.)

To make chip-bag confetti, harvest and reuse the circular by-products of your hole puncher.

SCIENCE JOURNAL

CHIPS

→ INDUSTRY NAME: CHIPS
→ LATIN NAME: GLOBULI SOLANIANI (ACCORDING TO THE VATICAN)
→ MADE FROM: POTATOES, SALT ÷ OIL (PLUS FLAVORINGS)

DIMENSIONAL CHARACTERISTICS

TECHNICAL NAME: HYPERBOLIC PARABOLOID!

KRINKLE (AND THICKNESSES) $\frac{}{}$ 210/1000"

REGULAR $\frac{}{}$ 55/1000"

KETTLE $\frac{}{}$ 98/1000"

BAKED $\frac{}{}$ 41/1000"

SADDLE

$\frac{}{}$ 50/1000"

MADE FROM MASHED UP SPUDS

$$z = a\left(x^2 - y^2\right)$$ WHERE a IS A CONSTANT > .01

LIDS

GATE

AVG. LID THICKNESS = 22/1000 INCHES

→ INDUSTRY NAME: OVERCAP LID
→ SCIENTIFIC NAME: LINEAR LOW-DENSITY POLYETHYLENE (LLDPE)
→ MADE FROM: PLASTIC

---

14 *The Chipping News*

Working with Chip Lids: The Hole Story

To poke a small hole: Depends on lid. Not all lids are made from the same plastic. Opaque lids (for example, the yellow ones used by the producers of Stax) can generally be pierced with a pushpin or a nail. Transparent lids (for instance, the ones that come off Pringles tubes) tend to split when pierced, so it's better to ask an adult to puncture them with a hot pin.

To punch a ¼-inch hole: A decent hole puncher (and a little effort) works well on the perimeter of lids. Center holes are a little trickier.

You may have to ask a grown-up to use a grommet punch or a drill bit. If no adult is available, consider the following procedure:

1. Layer two or three soft rags flatly on a stable surface.

Grommet Punch

2. Place the lid on top of the rags and pierce it with a ballpoint pen. Once the starter hole is formed, you can enlarge its diameter by pushing the pen through the lid.

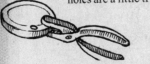

SPUDS

→ INDUSTRY NAME: POTATOES
→ LATIN NAME: SOLANUM TUBEROSUM
→ MADE FROM: STARCH, MOSTLY

KINGDOM → PLANTAE
DIVISION → MAGNOLIOPHYTA
CLASS → MAGNOLIOPSIDA
SUBCLASS → ASTERIDAE
ORDER → SOLANALES
FAMILY → SOLANACEAE
GENUS → SOLANUM
SPECIES → S. TUBEROSUM

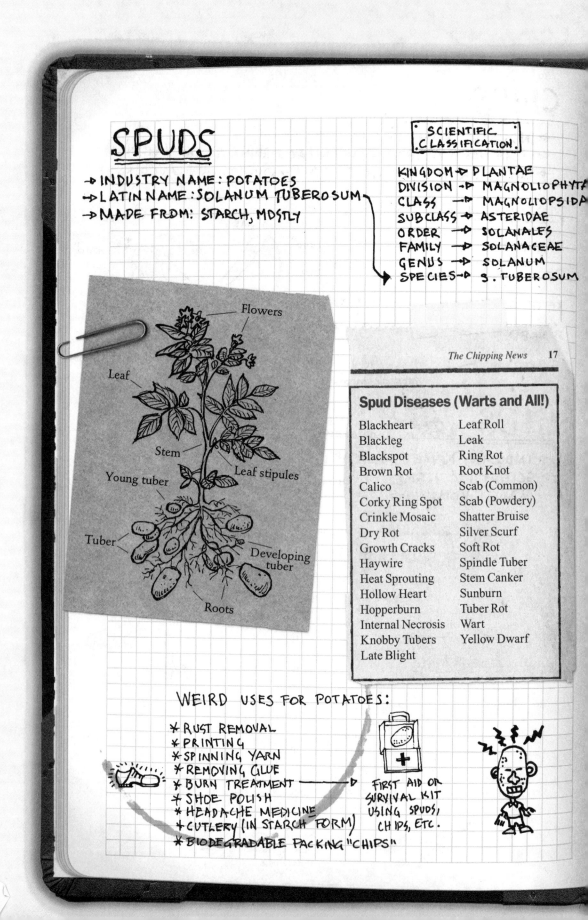

Flowers

Leaf

Stem

Leaf stipules

Young tuber

Tuber

Developing tuber

Roots

The Chipping News 17

Spud Diseases (Warts and All!)

Blackheart	Leaf Roll
Blackleg	Leak
Blackspot	Ring Rot
Brown Rot	Root Knot
Calico	Scab (Common)
Corky Ring Spot	Scab (Powdery)
Crinkle Mosaic	Shatter Bruise
Dry Rot	Silver Scurf
Growth Cracks	Soft Rot
Haywire	Spindle Tuber
Heat Sprouting	Stem Canker
Hollow Heart	Sunburn
Hopperburn	Tuber Rot
Internal Necrosis	Wart
Knobby Tubers	Yellow Dwarf
Late Blight	

WEIRD USES FOR POTATOES:

* RUST REMOVAL
* PRINTING
* SPINNING YARN
* REMOVING GLUE
* BURN TREATMENT ——▷ FIRST AID OR SURVIVAL KIT USING SPUDS, CHIPS, ETC.
* SHOE POLISH
* HEADACHE MEDICINE
* CUTLERY (IN STARCH FORM)
* BIODEGRADABLE PACKING "CHIPS"

TUBES

→ INDUSTRY NAME: RIGID CANISTER
→ SCIENTIFIC NAME: RIGHT CIRCULAR CYLINDER (HYPERBOLOID)
→ MADE FROM: HIGH-DENSITY POLYETHYLENE (HDPE)
   OR CARDBOARD & FOIL WITH METAL BOTTOM

Working with Tubes

Cutting off the bottom of a *plastic* chip tube is best done by an adult using a hacksaw. A bread knife or craft knife slices right through *cardboard* tubes.

Here's a way to cut through a cardboard tube that's safer for kids (though adult supervision is still required):

1. Mark the spot you want to cut off the tube by ringing it with a piece of masking tape.

2. Using a pushpin, poke holes closely together along the tape edge that marks the cutting line.

3. Push a picnic knife (preferably one made from potato starch) from hole to hole until you separate the bottom of the tube.

VOLUME = $\pi r^2 h$

h

SURFACE AREA = $2\pi r (r+h)$

TO DO
- ☑ READ LEON & THE CHAMPION CHIP
- ☐ TEST PROPULSION PIPE (MINI-MARSHMALLOWS?)
- ☑ MIX UP BATCH OF SPUD CRUD
  YUCK!

The Chipping News

Tube Puncturing

To puncture the bottom of a chip tube, a hammer and thin nail usually does the trick.

www.potatochipscience.com
Before Visiting
After Visiting
www.potatochipscience.com

BAGS

Never **throw chip bags away. Here's why:**
They can float your boat. If you're ever lost, they can
help point you in the right direction. Chip bags can
even be electrifying. Don't believe us? Then go fly a kite!
(Using a chip bag of course.)*

*Instructions can be found on page 10.

The potato was the first farm crop grown in outer space.

BAG BLASTER

Build an Air Rocket Out of Potato Chip Bags

FIELD: ROCKETRY
CONCEPT: PROPULSION

MATERIALS

Scissors

2 medium-size (2.5-ounce) chip bags

Paper towel

Bendy straw

Transparent tape

METHOD
To Make the Rocket

1 Cut open one empty chip bag, wipe off the grease inside with the paper towel (page viii), and cut a foil rectangle roughly 9×5 inches.

2 Place the bendy straw against a long edge of the rectangle. Roll the foil around the straw and secure the edge with tape. Remove the straw and set it aside.

3 Complete your rocket by taping shut one end of the foil tube.

To Make the Launchpad

1 Tape the short portion of your bendy straw inside the second chip bag.

2 Tape the open seam of the chip bag closed. (Use plenty of tape to form an airtight seal.)

BAGS

TO LAUNCH THE ROCKET

❶ Inflate the launchpad by blowing into the exposed end of the straw. When the bag is inflated, pinch the straw closed with one hand to keep the launchpad puffy. (If the bag doesn't inflate, there's probably a leak. Cover any holes with tape.)

❷ Use your other hand to slip the rocket onto the exposed portion of the straw as you let go of the straw. When you're ready, start the countdown ("3…2…1…") and… SQUEEZE the launchpad! Be sure to aim your rocket away from humans and pets.

If the launchpad springs a leak, just repair it with tape.

BLAST OFF!

MUNCH ON THIS

Dart flight fins   Paper clip nose cone   File folder tab fins

To add stability, you can modify your rocket with dart flights, paper clips, and file folder tabs.

Continued on next page

MEANING

All rockets—whether they travel to the moon or barely reach the kitchen ceiling—are subject to three forces during flight: thrust, drag, and weight (gravity).

Thrust is the aerodynamic force that launches the rocket. When the inflated chip bag is squeezed, air pressure shoots out through the bendy straw and toward the closed-off end of the rocket. The pressure inside the hollow rocket tube creates the *oomph!* (technically known as thrust) needed for a successful blastoff.

Another aerodynamic force known as drag also comes into play. Drag is the resistance in the atmosphere that slows the rocket's flight.

One *non*aerodynamic force additionally affects the flight of a chip bag rocket: weight, the gravitational force that pulls the homemade missile (and everything else that stays within the Earth's atmosphere) back to the ground.

MUNCH ON THIS

According to NASA, bags of potato chips on super-modified jets have been known to explode soon after takeoff. The reason? The sudden change in air pressure.

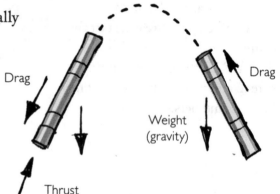

Drag

Drag

Weight (gravity)

Thrust

Squeezing the launchpad pushes air through the straw, creating the thrust that sends the rocket skyward. Drag slows down the rocket, and gravity brings it back to Earth.

Three Forces of Foil-Tube Flight

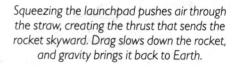

Thrust	The force that moves an object forward.
Drag	The force that slows a solid object (a rocket, for instance) as it moves through a liquid or a gas.
Weight	The measurement of the gravitational for ce that acts on an object.

COMPASS

FIELD: **NAVIGATION**
CONCEPT: **MAGNETISM**

Lost? This Survival Aid Foils Disaster

BAGS

MATERIALS

| Chip bag (any size) | Scissors | Paper towel | Chip lid filled with water | Sewing needle | Magnet |

METHOD

① Cut open the empty chip bag and wipe off the grease inside with the paper towel (page viii).

② Cut a ½-inch square from the chip bag.

③ Fill a chip lid with water and float the foil square on its surface.

④ Stroke the sewing needle with the magnet 30 times. Make sure you hold the needle by the eye, and stroke only in one direction—toward the point.

⑤ Gently place the sewing needle on the foil square and watch the point aim northward.

MEANING

Your compass might be lightweight, but it points out some pretty heavyweight science. The magnetized needle floating on the foil swings to the north because it reacts to a much, *much* bigger magnet—a massive ball of molten iron buried deep inside the Earth. The electrical currents inside that ball of metal produce a magnetic field powerful enough to control the motion of compass needles.

To understand why the compass points north and not, say, east, it's helpful to imagine that the planet's core contains a huge bar magnet that runs from the North Pole to the South Pole. Because the opposite ends of magnets always attract (lining up "head to tail"), the magnetized tip of the needle is drawn to its polar opposite.

The Geographic North Pole is a magnetic south pole.

For more survival gear, turn to page 46.

CHIP-SHIP CHALLENGE

How Many Pennies Can You Float Inside a Chip Bag?

FIELD: HYDRODYNAMICS
CONCEPT: BUOYANCY

MATERIALS

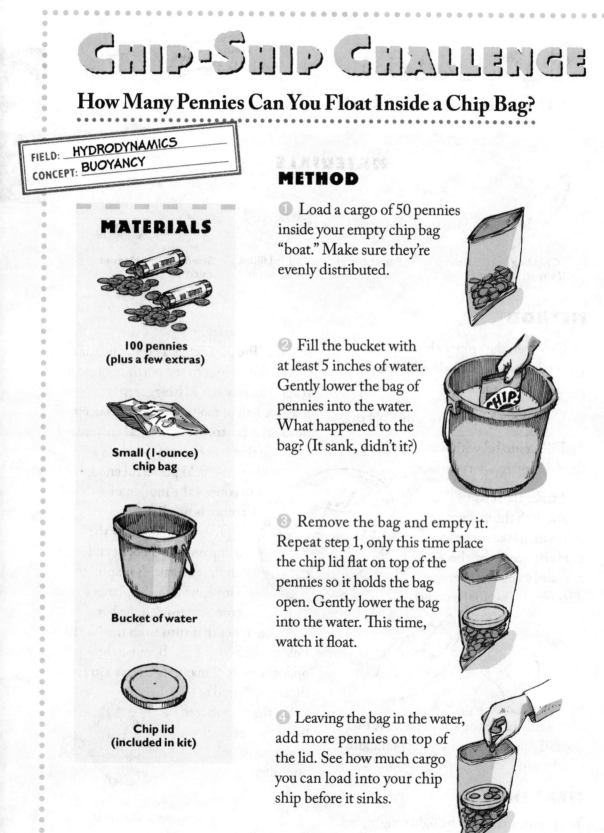

**100 pennies
(plus a few extras)**

**Small (1-ounce)
chip bag**

Bucket of water

**Chip lid
(included in kit)**

METHOD

① Load a cargo of 50 pennies inside your empty chip bag "boat." Make sure they're evenly distributed.

② Fill the bucket with at least 5 inches of water. Gently lower the bag of pennies into the water. What happened to the bag? (It sank, didn't it?)

③ Remove the bag and empty it. Repeat step 1, only this time place the chip lid flat on top of the pennies so it holds the bag open. Gently lower the bag into the water. This time, watch it float.

④ Leaving the bag in the water, add more pennies on top of the lid. See how much cargo you can load into your chip ship before it sinks.

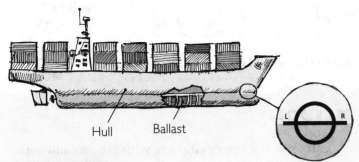

All oceangoing cargo ships display a "load line" on their hulls—a circle with a horizontal bar cutting through it. If the bar drops beneath the water, the crew knows the ship is overloaded.

Hull   Ballast

MEANING

The chip ship, like all cargo vessels, is designed to carry as much weight as possible. What keeps a boat from sinking? Buoyancy—the upward force that liquids and gases exert on objects.

The more water the hull of the ship displaces (pushes aside), the more buoyant the vessel. The more buoyant the vessel, the more cargo it can carry.

Positioning a lid on top of the pennies broadens the hull of your chip ship, which increases the volume inside the vessel and the amount of water displaced outside. The pennies at the bottom of the chip ship also serve as ballast—weight that improves stability. If properly designed, your chip ship can carry cargo more than 100 times its own weight.

CHIP CHALLENGE

Can you make a chip ship that holds more than 100 pennies? (Hint: Try keeping the bag open with cotton swabs or toothpicks.)

This trick really floats my boat.

Boats (and bags) with broad, stable hulls do a fine job of carrying cargo.

ELECTRIC WAVE

Static Lifts Sports Fans off the Edge of Their Seats

FIELD: PHYSICS
CONCEPT: ELECTROSTATICS

MATERIALS

Chip bag (any size)

Scissors

Paper towel

Marker

Transparent tape

Balloon

Head of hair

METHOD

1 Cut open the empty chip bag with scissors and wipe off the grease inside with the paper towel (page viii).

2 Cut a square (any size) from the chip bag, fold it in two, and, using a marker, outline *half* a person, with the fold down the middle.

3 Cut along the mark, open up the foil, and inspect. Voilà! A "sports fan."

4 Tape your sports fan to a table edge (or seat) so that he or she hangs over the edge.

5 Blow up a balloon and tie it closed. Rub the balloon vigorously on your hair at least 10 times.

6 Lower the balloon toward your sports fan so that it comes close but doesn't touch. Can you make your foil fanatic do the (static) electric wave?

Sports fan

MEANING

Have you ever gotten a nasty shock by touching a doorknob on a dry day in wintertime? The cause is static electricity. A sports fan fashioned from a chip bag provides a painless demonstration of this scientific phenomenon. By rubbing a balloon against your hair, you transfer negatively charged particles called electrons onto the balloon's surface. When the balloon is brought close to the fan, the "cheering" begins because the foil contains a larger ratio of *positively* charged particles called protons. Opposites attract. The balloon, loaded up with negative charges, draws the positively charged foil fan toward it.

Static electricity makes these foil fans do The Wave.

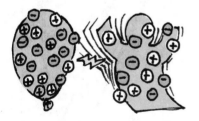

The negative charges on the balloon attract the positive charges on the foil fan.

Once the balloon and foil fan touch, the imbalance between the two materials is discharged (it disappears) and the cheering dies down—until you charge up the balloon again.

CHIP CHALLENGE

Can you coax a static reaction from a Ping-Pong ball? Potato chip crumbs? What about pepper or pieces of chip-bag confetti?

MUNCH ON THIS

Static shocks are uncommon in damp tropical environments because humidity neutralizes electrical imbalances.

BAGS

KITE

FIELD: ENGINEERING
CONCEPT: AERODYNAMICS

We Call This Low-Cost High-Flyer the *Chipsqueak*

METHOD

➊ Cut open the empty chip bag with scissors and wipe off the grease inside with the paper towel (page viii).

➋ Make the spar (the horizontal crosspiece) by using scissors to cut down the first coffee stirrer to 3 inches. Mark the midpoint (1½ inches) of the spar with an *X*.

1½"

3"

➌ Using the single-sided tape, attach the spar to the middle of the bag.

➍ Make the kite's spine (the vertical support) by cutting down the second coffee stirrer to 4 inches. Using the pushpin, prick a hole in it 1 inch from the top.

4"

1"

➎ Feed the string through the pinhole in the spine. Tie a knot at the end of the string to keep it from slipping back through the hole. (If you're using fishing line, you may need to tape the knot to the spine.)

Spar

Spine

Tail

MATERIALS

Scissors

Medium-size (2.5-ounce) chip bag

Paper towel

2 wooden coffee stirrers

Marker

BAGS

⑥ Place the spine across the spar so that the pinhole is over the *X* on the crosspiece.

⑦ With the single-sided tape, tape the top and bottom of the spine to the bag. (Important: The two pieces of wood should not be taped to each other.)

⑧ Run the double-sided tape between the ends of the spine and the spar in the form of a diamond.

⑨ Trim the bag all along the outer edge of the tape.

⑩ Notch the four corners of the kite, and remove the excess triangles of bag material.

⑪ Fold each taped edge of the bag over on itself.

⑫ Tie the loose end of the string to the Popsicle stick, tape on a tail made from a foil scrap, and head outside on a windy day.

You can also use this design to make a large kite. Simply maintain the 4:3 length ratio between the spine and the spar.

Continued on next page

Transparent tape (single-sided)   **Pushpin**   **Kite string or fishing line**   **Transparent tape (double-sided)**   **Popsicle stick**

Parachutes are basically anti-kites. (They're designed to float down instead of up.)

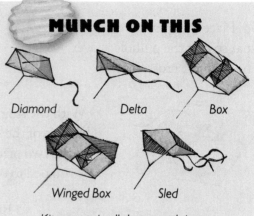

MUNCH ON THIS

Diamond   *Delta*   *Box*

Winged Box   *Sled*

Kites come in all shapes and sizes.

MEANING

Ever since the Chinese started turning bamboo, silk fabric, and thread into fluttery flying machines some 3,000 years ago, scientists and backyard inventors have loved designing kites. Both Alexander Graham Bell and the Wright brothers experimented with various shapes

Alexander Graham Bell

and materials in the hope of lifting humans off the ground. Benjamin Franklin, more famously, flew a kite to test his theories of static electricity. Recent designers have built kites smaller than a penny and larger than half a football field.

Kites have been used to monitor wind strength, measure distances, and send signals. They've been used to pull boats, buggies, and snowboarders. Fishermen sometimes use kites to catch fish.

Yet regardless of their shape, size, or function, all kites are subject to the same physical forces. Every airborne kite confirms both Newton's Third Law of Motion, which states that for every action there is an equal and opposite reaction,

and Bernoulli's Principle, which states that as the speed of the flow of a fluid such as air—yup, physicists consider air to be a fluid—increases, the pressure in that fluid decreases.

Put more simply, a kite stays up (or crashes) because of the amount of air pressure exerted on it (that's where Newton comes in) and because of the speed of the airflow that passes over and under the wing (Bernoulli's insight). If air moves more slowly on the underside of the kite than on the top, the air pressure underneath will create lift. But if the air pressure is greater on the top of the kite, drag is created and it will come tumbling back to Earth. (See page 37 for more about Daniel Bernoulli.)

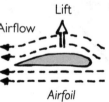

Lift

Airflow

Airfoil

HYDROFOIL

Dish Soap Gives This Chip Ship Zip

BAGS

MATERIALS

Scissors | **Chip bag (any size)** | **Bucket of water** | **Toothpick** | **Dish soap**

METHOD

① Cut a small triangle from an empty chip bag. (This will be your chip ship.)

② Put some water in the bucket and position your ship at the edge of the "lake" so that it points inward.

③ Dip one end of the toothpick in a drop of dish soap, then touch the toothpick to the water just behind the ship.

④ Watch your chip ship zip! (If you want to repeat the experiment, you'll have to change the water.)

MEANING

What gives your hydrofoil its zip? The answer can be found in the special nature of water. Unlike many liquids—alcohol, for instance—water molecules tend to stick together. This clinging action, called cohesion, produces surface tension that provides water with an "invisible skin."

Surface tension explains why drops of dew collect on leaves and why water striders can shoot across the top of a pond. In fact, the cohesive force of water is so strong it can hold a paper clip gingerly placed on its surface. (Try it!)

However, when a drop of dish soap is added to water, the surface tension is reduced. Soap molecules puncture the water's skin, triggering a molecular wave that pushes the ship forward.

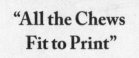

THE CHIPP

Serving Potato Chip

Vol. 1, No. 1

Potato Chip–Bag Technology Adapted to Solar Science

[Chipping News Network—CNN] It's been used to decorate T-shirts, protect rare comic books, and improve the design of hang gliders. Now companies around

Flexible film has been used in chip bags, space suits, and hang gliders. Are solar cells next?

the world are racing to turn flexible film packaging—the laminated plastic-and-metal material commonly associated with potato chip bags—into semiconductor solar cells. The cutting-edge technology involves applying a super-thin mixture of gallium, indium, copper, and selenide onto rolls of the film. Known as GICS, the solar film is not yet as efficient as the silicon panels often used in sunny regions throughout the world, but efforts to improve energy conversion rates might soon enable this innovation to be integrated into roof tiles and other building materials.

March 14 Marks Two Industry Milestones

[Chipping News Network—CNN] March 14 is not just National Potato Chip Day. It also marks the birth of Albert Einstein. Although principally known for his insights into physics, the Nobel laureate also theorized about a saddle-shaped universe half a century before the birth of the saddle-shaped potato chip. Coincidence?

Albert Einstein

ASTROPHYSICISTS TO SCHOOL BOARD: HEY, IT *IS* ROCKET SCIENCE!

[Chipping News Network—CNN] Arizona seventh-grader David Silverstein, inspired by the movie October Sky, took a homemade rocket made from a potato chip tube to school. After one look at the clever missile, administrators determined it was inappropriate. They confiscated the rocket and suspended its inventor. News of the harsh punishment quickly spread throughout the country, and ultimately prompted organizers of an annual aeronautics convention to honor David as a special guest.

ING NEWS

Science Since 2009

March 14, 2009

SPECIAL INVESTIGATION

NEWS SOURCE BURSTS BAG INDUSTRY SECRET

UNCOVERS TRILLION-INCH PACKAGING STAT

By CHIP ENDALE
Senior Correspondent

[Chipping News Network—CNN] Makers of potato chip bags rarely discuss how much film material they produce. But now a senior-level potato chip executive has broken the silence, providing *The Chipping News* with bag production figures long kept under wraps.

"We use about 15 billion square inches of packaging

film a year," the source acknowledged, adding, "And we're about 1.5% of the total [U.S.] potato chip market."

Based on this revelation, and other statistics, *Chipping News* investigators estimate that the United States generates 1 trillion square inches of potato chip bags annually.

A top analyst who crunches numbers (and chips) at the Potato Chip Science packaging center put this statistic in

perspective: "If you combined all the chip bag material that the U.S. produces into a single humongous bag, you could wrap all of the Vatican, Liechtenstein, and San Marino. And you'd *still* have plenty of material left over to wrap every major city in America, from New York to L.A.! It's a pity we don't teach our kids how to reuse packaging. They could turn this so-called trash into experimental treasure!"

For more, go to
www.potatochipscience.com

1,000,000,000,000 square inches of bag material could cover...

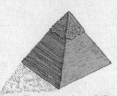

Singapore   600,000 Statues of Liberty   7,500 Pyramids of Giza

POTATO CHIP SCIENCE

CHIPS

Soak it. Burn it. Crush it. Bake it.
But no matter how you slice it, the potato chip is the
world's most popular snack.

CHIP ANALYZER

AN ADULT MUST CHIP IN

What Do Car Batteries and Vinegar Chips Have in Common?

METHOD
TO PREPARE THE INDICATOR JUICE

FIELD: CHEMISTRY
CONCEPT: ACIDS & BASES

① Have an adult assistant chop up 1 cup of red cabbage and place it in the blender.

② Blend in small amounts of distilled water until the cabbage is pulped. (One cup of water should be plenty.)

③ Strain the pulp. The liquid that remains in the bowl is your indicator juice. (The pulp can be tossed or, better still, composted; see page 26.)

Pulp

Indicator juice

Unless refrigerated, indicator juice begins to stink after a few days.

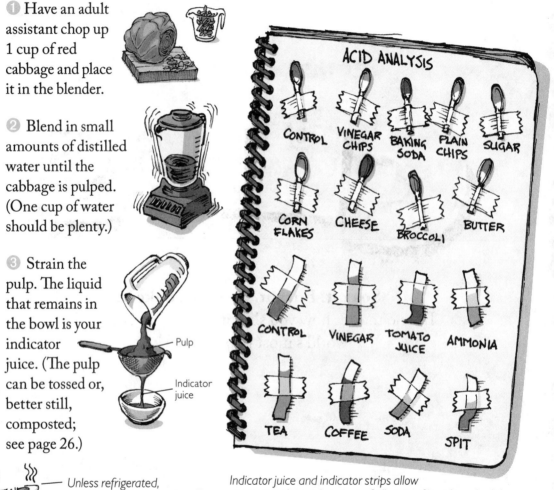

Indicator juice and indicator strips allow you to test the acid content of all sorts of liquids and solids.

MATERIALS

 I cup chopped red cabbage

 Blender

I cup distilled water

 Strainer and bowl

Cotton swabs

 Transparent tape

To Test the Acidity of Solids

1 Dip a cotton swab into the indicator juice. Once the swab is dry, tape it to a lab notebook or a sheet of paper and mark it "Control." (A control is an untested specimen against which other experimental samples can be compared.)

2 Place a tiny amount of a test solid— a vinegar chip, for example—into a clean chip lid.

3 With the straw, sprinkle a few drops of the indicator juice over the solid and observe whether the liquid changes color.

4 Record any color change by rubbing a fresh cotton swab against a moistened area of the chip. Once the swab is dry, tape it next to the control swab, and write down the name of the analyzed substance.

5 Repeat the test on other solids (salt, baking soda, all kinds of snack foods).

To Test the Acidity of Liquids

1 Dip strips of uncoated (nonshiny) paper into the indicator juice.

2 Dry the soaked indicator strips on a sheet of aluminum foil or wax paper. (Never place wet indicator strips on newspaper. The acid in the printer's ink can produce a false result.)

3 Before testing any liquid, record a control strip by taping and labeling an unused indicator strip on a sheet of paper or in a lab notebook.

4 Place a tiny amount of a test liquid—lemon juice, for example— into a clean chip lid.

5 Dip a new indicator strip into the test liquid and observe the results. Did the strip change color?

6 Repeat the test on other liquids (water, soda, ammonia, apple juice).

Continued on next page

CHIPS

Notebook or sheet of paper   **Chips (vinegar and plain) and additional test materials**   **4 chip lids (included in kit)**   **Straw**   **Strips of uncoated paper**   **Aluminum foil or wax paper**

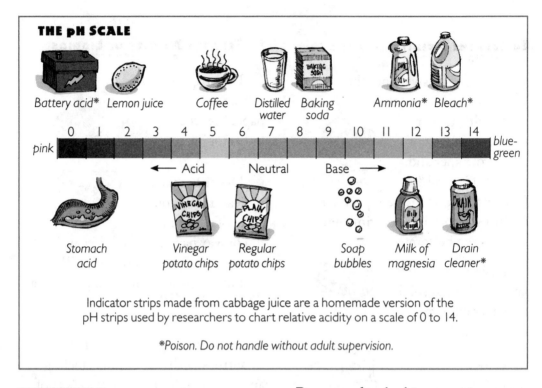

THE pH SCALE

Battery acid*   Lemon juice   Coffee   Distilled water   Baking soda   Ammonia*   Bleach*

0 1 2 3 4 5 6 7 8 9 10 11 12 13 14

pink   blue-green

← Acid   Neutral   Base →

Stomach acid   Vinegar potato chips   Regular potato chips   Soap bubbles   Milk of magnesia   Drain cleaner*

Indicator strips made from cabbage juice are a homemade version of the pH strips used by researchers to chart relative acidity on a scale of 0 to 14.

*Poison. Do not handle without adult supervision.

MEANING

Cabbage juice provides a simple way to indicate the relative acidity of a substance. The presence of acid will turn the indicator pink, whether it's in liquid or strip form. Vinegar chips contain enough acid to produce a color change, whereas plain chips do not.

The chemical opposite of an acid is called a base, or alkali. Basic substances will turn cabbage juice blue-green. Neutral substances—distilled water, for instance—are neither acid nor base, so they do not change the color of the indicator. When mixed together, acids and bases neutralize each other—their opposing properties cancel out.

Acids tend to be sour-tasting chemical compounds. Some acidic substances—vinegar chips, for example—are edible. Some are not edible. (Battery acid is *super* poisonous.)

Bases tend to be bitter-tasting compounds. Like acids, some bases can be eaten and some can't be. Egg whites are basic and make delicious meringues. Then again, drain cleaner, which is also basic, is *extremely* toxic.

How Acidic Is Acid Rain?

Until recently, rainwater pretty much always had a pH (the measure of acidity and alkalinity) of around 5.6. The carbon dioxide found naturally in the atmosphere made the rain slightly acidic. Nowadays, however, acid levels have increased greatly because of air pollution.

When we burn fossil fuels (to produce electricity or make our cars run), acidic gases are emitted into the air. Rain brings those toxic substances back to the ground. The result: acid rain.

Near some chemical plants, the pH level of rainwater has registered 2.5—the acid level of lemon juice and vinegar.

CSI* DETECTIVE KIT

Here's One Way to Finger(print) a Chip Thief

AN ADULT MUST CHIP IN

*Chip Science Institute

FIELD: **FORENSICS**
CONCEPT: **FINGERPRINT ANALYSIS**

METHOD
TO MAKE FINGERPRINT POWDER

1 Place a pie pan in the kitchen sink and a potato chip in the pie pan. Then ask an adult to burn the chip to a blackened crisp by lighting it with a match.

2 Once the chip has cooled, crush the charred remains into tiny flakes with the back of a spoon. Ta-dah! You've just made fingerprint powder.

3 Store your fingerprint powder in a freezer bag or a small jar.

Potato chips are HIGHLY flammable. Take special care when burning them.

CHIPS

MATERIALS

Metal pie pan

1 potato chip

Matches

Spoon

Freezer bag or small jar

Suspect

Fingerprint card (page 23)

Transparent tape (standard ¾-inch width)

Clear packing tape (standard 2-inch width)

Continued on next page

The Arch

Forms a bell-shaped curve

The Loop

Starts and ends at the same side

The Whorl

Roughly circular

Before computers, investigators relied on various nondigital systems of fingerprint classification. One such method—introduced by Sir Edward Henry to Scotland Yard in 1901—classified a number of basic fingerprint patterns, including the arch, loop, and whorl.

TO TAKE PRINTS (1-FINGER METHOD)

1 Have the suspect rub a pinch of fingerprint powder between his or her thumb and index finger.

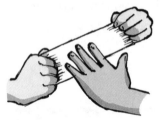

2 Have the suspect press the blackened index finger against the sticky side of a piece of transparent tape. Apply the tape to the appropriate box of the fingerprint card.

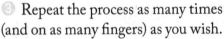
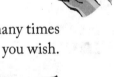

3 Repeat the process as many times (and on as many fingers) as you wish. Compare the prints to the patterns in the illustration above.

TO TAKE PRINTS (5-FINGER METHOD)

1 Have the suspect rub fingerprint powder between all five fingertips of one hand.

2 Hold at least 6 inches of clear packing tape, sticky side out.

3 Have the suspect press the pads of the five blackened fingertips against the sticky side of the tape. Then apply the tape to the fingerprint card.

According to the FBI, if a person has more than ten fingers, only the thumbs and the next four fingers of each hand should be printed. All extra digits should be ignored.

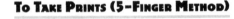

MEANING

The skin on the human fingertip, the part technically known as the bulb but better known as the pad, is covered with tiny hills and valleys called friction ridges. These ridges retain dirt, grime, and oils (potato chip grease being a prime example) that leave behind fingerprints—unique, telltale marks. Dactyloscopists (fingerprint experts—*dactyl* means "finger") typically work with three kinds of prints. The most common is the latent print, an invisible trace of human contact that requires special fingerprint powder and a dusting brush to lift. More unusual are visible prints—the marks left behind when a finger touches ink, blood, or some other trace-producing substance. On rare occasions, a detective might come across a "plastic" print—a three-dimensional impression left in a soft medium such as margarine, dust, soap, or even Silly Putty.

Once any one of these three kinds of prints is "lifted" (recorded), it can be compared to digital records in various criminal databases. Fingerprint analysis solves many more "unknown suspect" cases than DNA tests do.

MUNCH ON THIS

Koala fingerprints look so humanlike it's tough to tell the two apart, even with an electron microscope.

C.S.I.
(Chip Science Institute)

NAME : _____

AGE : _____

FAVORITE POTATO CHIP : _____

FINGERPRINT CARD

1. RUB FINGERPRINT POWDER BETWEEN FINGER(S) WITH GRINDING MOTION.
2. PRESS BLACKENED FINGER(S) AGAINST TAPE. DO NOT "ROLL" FINGER(S).
3. APPLY TAPE TO FINGERPRINT CARD.

LEFT LITTLE	LEFT RING	LEFT MIDDLE	LEFT INDEX	LEFT THUMB
RIGHT THUMB	RIGHT INDEX	RIGHT MIDDLE	RIGHT RING	RIGHT LITTLE

FORM NO. PCS - 070312

Copy this fingerprint card to create files on friends, family, teachers, and other potato chip scientists.

DANCING CHIPS

An Acoustical Drum with a Chipper Beat

MATERIALS

2 empty chip tubes without lids (metal-bottomed tubes work best)

Plastic wrap (or a big balloon with its neck cut off)

Rubber band

1 potato chip

Wooden spoon

METHOD

1. Cover the open mouth of the first tube with plastic wrap, and secure it with a rubber band.

2. Gently pull down on the plastic wrap to tighten the skin of the "drum."

3. Crush the potato chip and place the crumbs on the drum.

4. Hold the mouth of the second tube close to the crumbs. With your free hand, bang the wooden spoon on the bottom and watch the chip crumbs dance.

Drum skin

If you watch the drum from the side, you'll get a better view of the "dancing."

MEANING

What makes the chips dance? Sound waves.

Banging on the open tube triggers the movement of air particles, which bounce into other air particles in an expanding wave pattern that has a physical impact. When those invisible waves reach the drum, they make it vibrate, which in turn causes the chips to jiggle and jump.

Ear drum

The human ear works much like the chip drum does. Air molecules bounce against the surface of the eardrum, causing it to vibrate. Those vibrations are turned into nerve impulses that travel to the brain, where they register as sound.

The harder you bang on the drum, the bigger the waves. Scientists often translate the intensity of these waves into a system of relative sound measurements known as decibels (dB).

CHIPS

Measuring Sound From *Crinkle* to *Kerplouie!*	Decibels (dB)
Whispering (about potato chips, for instance)*	30dB
Stale potato chip (first crunch, closed mouth)*	50dB
Fresh potato chip (first crunch, closed mouth)*	55dB
Fresh potato chip (first crunch, open mouth)*	63dB
Two kids talking (about potato chips, for instance)*	60–70dB
Average noise on a potato chip factory assembly line	75dB
City traffic (inside a potato chip truck, for instance)	85dB
Motorcycle	110dB
Rock concert (in front of speakers)	115dB
Jet engine at 200 feet	140dB

*Measurements taken from a distance of 20 centimeters.

CHIP CHALLENGE

How far apart can you keep the tubes and still make the crumbs dance?

At 190 decibels, the human eardrum explodes!

FIELD: ENVIRONMENTAL SCIENCE
CONCEPT: MICROBIOLOGY

COMPOSTER

Spud, Chip, or Bag: Which Sticks Around the Longest?

MATERIALS

1 small chip tube or other small container

Potting soil or a peat moss pellet

Water

1 potato chip

Potato knife (included in kit)

Potato (just the peel)

1 pinch chip-bag confetti (page viii)

METHOD

1 Fill the tube with potting soil and moisten it with water. (Dump any standing water.)

2 Crumble a potato chip and mix it into the soil.

3 Using your potato knife, peel an inch of skin off a potato, chop the skin into small pieces, and add the scraps to the soil.

4 Add the chip-bag confetti to the soil.

5 Once a day, for seven days, stir the contents of the container, occasionally adding water to keep things moist.

Keep the soil as damp as a wet sponge, but do not overwater.

6 Spud. Chip. Bag. Then watch how each disintegrates. Keep a record of the changes that take place during the week.

MEANING

All organisms (living things), whether plant or animal, eventually decompose (rot). Much of this decomposed matter can be turned into compost—nutrient-rich materials such as old eggs, cow poop, uneaten potato chips (if such things exist), and potato peels, mixed together with dirt and leaves. Compost not only fertilizes soil, helps sandy earth retain water, and keeps plant roots healthy, it also reduces the amount of garbage that's carted off to the dump.

What turns a potato peel into the gardener's best friend? Microorganisms. (*Micro* means "very tiny.") A single gram of compost can contain 2 billion invisible creatures that snack on the soil. Think of them as a backyard digestion team.

Inorganic matter (nonliving things)—chip-bag confetti, for instance—is not biodegradable and cannot be composted. Like most plastics, a chip bag would take decades, or even centuries, to disintegrate (break down). That's one reason potato chip trash, along with other forms of packaging, makes up to one third of the quarter trillion tons of waste Americans produce each year. On the other hand, biodegradable materials such as potato peels, potato chips, and potato knives have a relatively rapid rate of decomposition.

Bacteria make up 80 to 90% of the microorganisms in compost. Fungi and other microscopic organisms make up the rest.

CHIPS

Decomposition Rates (A Highly Selective List)

Object	Material	Decomposition Rate
Potato Peel	Starch	2–10 days
Potato Chip	Potato, salt, oil	20–50 days
Potato Knife	Potato starch	90–180 days
Chip Bag	Laminated plastic and metal	10–80 years
Plastic Knife	Polypropylene	Up to 250 years

MUNCH ON THIS

Every year, more than 10 billion picnic knives end up in America's landfills. Those knives can take centuries to decompose unless they're made out of potato or corn starch. Often molded from waste produced at chip factories (page 66), starch knives are more heat resistant than their plastic cousins and, under the right conditions, decompose in just 6 months. You can add potato knives to your compost and then use that compost to grow spuds.

Here's one method to do just that: Lay down a tire, fill it with composted soil, and plant several sprouted potatoes. (Even pieces of potatoes with eyes will work.) When plant leaves reach the top of the first tire, add another tire and enough soil to cover all but the topmost leaves.

Tire planter

Energy=Munching Chips, Too!

E=MC²

AN ADULT MUST CHIP IN

Where Snackin' Meets Calculatin'

FIELD: __NUTRITION__

CONCEPT: __FOOD ENERGY__

CHIP CHALLENGE

What is the caloric value of a single Potato Chip Crunchie? (Answer is at the bottom of this page.)

LAB WORK

Leon's Potato Chip Crunchies

MATERIALS

½ cup sugar

1 stick butter

1 egg

1 teaspoon vanilla extract

1 cup all-purpose flour

2 1-ounce bags plain potato chips, slightly crushed

METHOD

Preheat the oven to 350 degrees Fahrenheit

1. Mix the sugar and the butter until smooth.

2. Mix in the egg.

3. Mix in the vanilla.

4. Mix in the flour.

5. Add the chip crumbs to the mix.

6. Divide the dough into 12 globs spaced evenly on a large ungreased cookie sheet.

7. Bake for 5–8 minutes.

Makes 12 Crunchies

DATA ANALYSIS

Once you and your adult assistant complete the lab work in the kitchen, you can tackle the chip challenge above. Here's how:

❶ Calculate the calories contained in a whole batch of Crunchies by adding up the caloric values of the ingredients:

½ cup sugar = 387 calories

1 stick butter = 800 calories

1 egg = 70 calories

1 teaspoon vanilla = 2 calories

1 cup flour = 440 calories

2 1-ounce bags potato chips = 300 calories

❷ Divide the total number of calories of all ingredients (1,999) by the number of Crunchies contained in the batch (12).

When conducting a snacktivity (or any other experiment) never taste research material unless the method states to do so.

There are at least two ways to prove that a Crunchie is packed with energy: you can burn it (*not* recommended) or you can eat it (highly recommended). Once you've crunched the numbers, make sure you crunch the Crunchies!

MEANING

Food provides the body with fuel. The energy value of that fuel can be measured in calories. If you eat more calories than your body burns (by exercising, breathing, sleeping, conducting potato chip science experiments, etc.), your body stores the extra energy as fat. If you don't eat enough calories, that fat gets used up.

Not all calories are created equal. Calories found in snack foods such as cookies and potato chips (as well as cookies made with potato chips) may taste good, but they provide almost no nutrients for the body. Raw fruits, vegetables, lean meats, brown rice, and whole-grain bread, on the other hand, are all rich in the kinds of vitamins, minerals, and proteins that allow the human engine to run smoothly. The average 11- to- 14-year-old boy requires 2,500 calories per day, and the average 11- to- 14-year-old girl requires 2,200 calories daily. These numbers vary depending upon the size and activity level of the kid. Bigger "engines" require more fuel.

MUNCH ON THIS

A pound of potato chips contains about 2,240 calories, and more than 1.8 billion pounds of potato chips are produced annually worldwide. That means the world consumes roughly 4 trillion calories of chips—enough energy to light up New York City for 40 days.

... AND THIS

An 8-inch stalk of celery contains 6 calories. But chewing and digesting that stalk requires more than 6 calories. That means eating a stalk of celery actually burns more energy than is generated.

... AND THIS

A tiny shrew requires so much energy, it needs to eat the equivalent of its own body weight every day. Imagine if you had to do that too.

CHIPS

A dozen Crunchies—1,999 calories—have the same caloric value as the following: breakfast (cold cereal, a cup of orange juice, buttered toast) plus lunch (tuna sandwich, a cup of fat-free milk, apple), plus a snack (a cup of low-fat yogurt with raisins), plus dinner (roasted chicken breast, a baked potato, peas, a whole-wheat roll, and green salad).

CHIPATORIUM

The twenty-two objects on display in this Cabinet of Potato Chip Wonders represent only part of the holdings of the Museum of Potato Chip Science. For more information about these (and other) rarities, visit www.potatochipscience.com.

No.1 Yo-yo – Formed from two chip lids, a Lego wheel, a rod, and a pair of wing nuts ☕ **No.2** The "Valentine"– Extremely rare cordate (heart-shaped) potato chip pulled from a bag of Miss Vickie's potato chips ☕ **No.3** Golden Champion Chip Medallion – Highest civilian award granted by the All-State Potato Chip Association (ASPCA) ☕ **No.4** Mrs. Potato Head – First edition (1953), wife of Mr. Potato Head, mother of Brother Spud and Sister Yam ☕ **No. 5** Potato Starch Knife – Biodegradable alternative to plastic cutlery ☕ **No.6** Tin Toy –Humpty Dumpty potato chip delivery truck, circa 1960 ☕ **No.7** The "Chip-a-pult" – Chip tube and potato-starch spoon snack-food launcher ☕ **No.8** Duncan "Little Champ" Kitty Clover Potato Chips top ☕ **No.9** Animator – Chip lid "thaumatrope" wheel that simulates motion when spun ☕ **No.10** Advertisement – Promotional clip for Slim Gaillard's zany (misspelled) 1952 song "Potatoe Chips" ☕ **No.11** Space Magic Coin – 1964 Krunchees Potato Chip Company premium, chronicling

U.S. space history and featuring astronaut Alan B. Shepard and President Kennedy 🖐 **No.12** Helicopter – Wooden spool transmission, pencil mast, and chip lid rotor blades 🖐 **No.13** Shrunken Head – Desiccated potato treated with natron (an embalming salt) 🖐 **No.14** Pinwheel – Chip-bag blades, map pin spindle, and pencil shaft 🖐 **No.15** Potato Mill – Spinning toy employing a stick, a string, a chestnut, and a potato 🖐 **No.16** Glider – Chip-bag fuselage and birdlike dihedral wings reinforced with glue and sewing-

thread guy wires 🖐 **No.17** Sail Car – Chip tube lid wheels and foil bag sail 🖐 **No.18** Parachute – Ketchup cup, single-serve chip bag, and thread 🖐 **No.19** Indicator Juice –Red cabbage solution used to test the relative acidity of solids 🖐 **No.20** Saratoga Springs, N.Y. postcard – Generally recognized as the birthplace, in 1853, of potato chips 🖐 **No.21** Pin – 1965 Guy's Potato Chips pin, Boston Red Sox 🖐 **No.22** Potato Chip Bag – Glassine chip bag bearing frowning (or smiling) man, circa 1940.

LIDS

Slower than a speeding bullet—when airborne.
Less powerful than a locomotive—when carrying
cargo. Unable to clear tall bookcases in a single flick—
when improperly tossed. It's the potato chip lid!
The lightweight disc can spin, glide, whirl,
twirl, flip, and fly. Not bad for a disc of
plastic that weighs $\frac{1}{8}$ of an ounce.

FIELD: **PHYSICS**

CONCEPT: **CONSERVATION OF ENERGY**

CHIPMOBILE

An Engine Powered by a Renewable Energy Source—Your Breath

MATERIALS

Pushpin

4 chip lids (included in kit)

Pencil

Large cardboard food container

2 bamboo skewers

Balloon (the bigger, the better!)

Bendy straw

Twist tie

Masking tape

Lightweight cargo

METHOD
To Build the Chassis (Body)

1. Use the pushpin to poke a small hole through the center underside of a chip lid: This will be the first of the four wheels.

2. Position the lid (with the pin still inside) at the spot where you'd like your front wheel to go, then poke the pin through the wall of the container to make the first of the four axle holes. Remove the pushpin and set it aside.

First axle hole

3. Push the pointy end of the bamboo skewer through the first wheel until the wheel is positioned ½ inch from the non-pointy end.

½"

4. Enlarge the hole *slightly* with the point of a pencil. (Don't make the hole too big!)

5. Feed the pointy end of the skewer through the axle hole and, using the pushpin, poke the second axle hole at the point where the skewer hits the far wall of the container.

6. Enlarge that second axle hole with the pencil.

Second axle hole

7. Feed the skewer and lid (axle and wheel) through the two axle holes, poke a hole through the center of a second lid, and push the pointy end of the skewer through it. Two wheels and an axle will now be in place.

8. Repeat steps 1 through 7 to attach the other axle and wheels.

To Power the Chipmobile

1 Insert the short end of the bendy straw into the balloon and secure it with the twist tie.

2 Position the straw lengthwise across the top of the chipmobile and tape it at both ends.

3 Blow up the balloon and pinch the straw to keep the air in. Without letting go of the straw, place the chipmobile on a smooth, hard surface and load it up with the lightweight cargo.

4 Bend the straw upward so that the balloon doesn't touch the ground, and then release your grip. With a bit of luck your chipmobile will zoom away.

TROUBLESHOOTING YOUR CHIPMOBILE: A CHECKLIST

Friction, the rubbing action between two surfaces, is the enemy of the chipmobile's performance. This resistance slows things down and shortens the distance the vehicle can travel between fill-ups. For the best results, make sure …

✓ The axles move freely in the axle holes.

✓ The front and back axles line up parallel to each other.

✓ The bottom of the chassis (the container) clears the ground.

✓ The wheels don't rub against the chassis.

✓ The balloon doesn't rub against the ground or the wheels.

LIDS

With this balloon car (and your breath), you never have to worry about the ballooning cost of fuel.

Continued on next page

Used Auto Parts

Tissue box

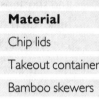
Chip tube

Takeout container

Function Name	Material	Alternate Materials
Wheels	Chip lids	Juice carton caps, CDs, thread spools
Chassis (body)	Takeout container	Tissue box, chip tube, packing foam
Axles	Bamboo skewers	Coat hanger wire, thin wooden dowels

MEANING

The chipmobile can't handle a lot of weight, but it does carry a full load of physics. For starters, it demonstrates the difference between potential energy and kinetic energy.

Potential energy, as the name suggests, is stored energy. An inflated balloon is a perfect example of potential energy. When you blow up a balloon, you're storing wind energy (your breath) inside. When you allow that energy to escape through the straw, the potential energy is transformed into kinetic energy—the energy of motion. The chipmobile also demonstrates the Third Law of Motion, which states that for every action there is an equal and opposite reaction. As the air whooshes out of the straw, the vehicle goes zooming off in the opposite direction. When the chipmobile finishes its journey, it confirms the First Law of Motion, which states that a body at rest stays at rest unless acted upon by an unbalanced force. Isaac Newton (1642–1727) first published the Laws of Motion in his *Principia* (1687).

Isaac Newton

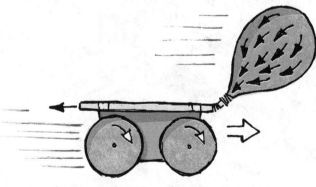
Air escapes the balloon in one direction and the chipmobile moves in the opposite direction.

AN ADULT MUST CHIP IN

SAUCER TOSSER

*DISC*over the Secret of Saucer Flight

MATERIALS

Chip lid
(included in kit)

Transparent
Tape

Craft knife

Glue

Lightweight objects

Wide-open space

METHOD

1. Look at the flying saucers on this page.
2. Allow your imagination to soar.
3. Design a saucer.
4. Toss it!

MEANING

For a saucer to soar smoothly, it needs both angular momentum (spin) for stability and the forward momentum caused by propulsion (a strong flick of the wrist) for distance. It stays in flight because of its shape, which allows air to flow faster over the top than under the bottom. The difference in flow speed creates a difference in air pressure, and the difference in air pressure produces lift.

This effect was first described by Daniel Bernoulli (1700–1782), a Dutch-born Swiss mathematician. (That's why it's named after him.) Bernoulli's Principle, which helps explain the influence of velocity (the combination of speed and direction) on the pressure of fluids (including air), has much more significant applications than causing a chip lid to fly. Airplane wings, Frisbees, and helicopter rotor blades are all designed using mathematical equations derived from Bernoulli's Principle.

Daniel Bernoulli

LIDS

SIGNAL MIRROR
A Flashy Desert Island Disc

AN ADULT MUST CHIP IN

METHOD

① Apply the black signal sticker to the inside of the chip lid.

② Cut a circle of foil from a chip bag or from a roll of aluminum foil to fit the lid and glue it on.

③ With an adult's assistance, use a craft knife to cut along the crosshairs that appear in the center of the signal sticker. The opening of the lines should be wide enough to let light through the lid.

④ To catch the attention of a rescue craft, follow the directions printed on the signal sticker.

Crosshairs

MEANING

An emergency signal mirror reflects a good deal about the physics of light. Rays of light generally bounce off flat (plane) mirrors the same way a billiard ball bounces off the side of a pool table. Here's how:

A ray of light (whether natural or artificial) that *approaches* a mirror is known as the incident ray. The ray of light that *leaves* the mirror is known as the reflected ray. The angle of the incident ray is always equal to the angle of the reflected ray. Physicists refer to this predictable behavior of light as the law of reflection.

MATERIALS

Black signal sticker (included in kit)

Chip lid (included in kit)

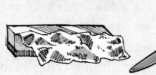
Aluminum foil

Scissors

Glue stick

Craft knife

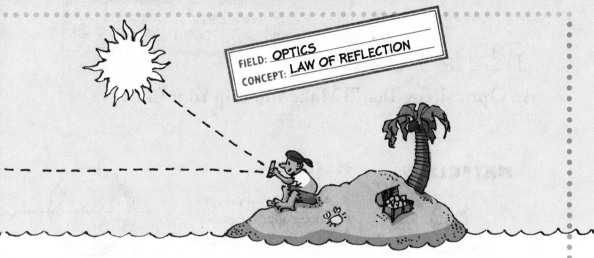

FIELD: OPTICS
CONCEPT: LAW OF REFLECTION

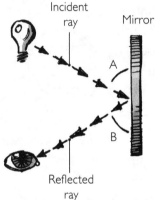

Incident ray

Mirror

A

B

Reflected ray

The law of reflection states that when a ray of light reflects off a surface (a mirror, for example) the angle of incidence (marked "A") equals the angle of reflection (marked "B").

MUNCH ON THIS

Back in the 19th century, armies employed a complex contraption called a heliograph that used mirrors and sunlight to flash coded messages over long distances. Only people in direct sight line of the device could see the signals.

LIDS

EMERGENCY SIGNAL MIRROR

1) HOLD THIS SIDE OF MIRROR CLOSE TO FACE AND REFLECT SUNLIGHT ON ANY NEARBY SURFACE IN LINE WITH TARGET.

2) TILT MIRROR TO SHINE AT TARGET.

3) TO SEND SIGNAL, LOOK THROUGH CROSSHAIRS AT TARGET AND JIGGLE MIRROR.

4) WHEN RESCUE CRAFT ARE NOT VISIBLE, AIM CROSSHAIRS AT A SPOT ALONG HORIZON.

This design is adapted from a signal mirror commonly used by military personnel during World War II. The effective range of a vintage WW II signal mirror is as much as 100 miles (when signaling airplanes). A chip-lid mirror can't manage that distance, but it's still pretty powerful.

FLIPPER

An Optical Toy That'll Make You Flip Your Lid

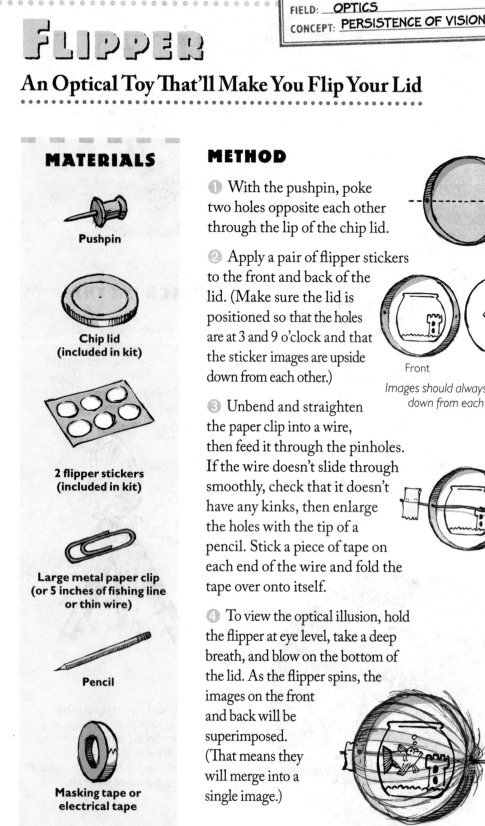

MATERIALS

Pushpin

**Chip lid
(included in kit)**

**2 flipper stickers
(included in kit)**

**Large metal paper clip
(or 5 inches of fishing line
or thin wire)**

Pencil

**Masking tape or
electrical tape**

METHOD

① With the pushpin, poke two holes opposite each other through the lip of the chip lid.

② Apply a pair of flipper stickers to the front and back of the lid. (Make sure the lid is positioned so that the holes are at 3 and 9 o'clock and that the sticker images are upside down from each other.)

Front   Back

Images should always be upside down from each other.

③ Unbend and straighten the paper clip into a wire, then feed it through the pinholes. If the wire doesn't slide through smoothly, check that it doesn't have any kinks, then enlarge the holes with the tip of a pencil. Stick a piece of tape on each end of the wire and fold the tape over onto itself.

④ To view the optical illusion, hold the flipper at eye level, take a deep breath, and blow on the bottom of the lid. As the flipper spins, the images on the front and back will be superimposed. (That means they will merge into a single image.)

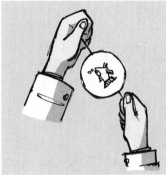

FLIP TIPS

● You can position flipper sticker designs so they superimpose on each other (for example, a fish in a fish bowl), or draw images so that they appear next to each other.

● If the image appears upside down while you're flipping, give the lid a half turn to make the image appear right-side up.

CHIP CHALLENGE

Can you create your own flipper designs on blank stickers?

MEANING

Your eyes and brain retain an image for a split second even after that image disappears. (If that didn't happen, the world would go dark every time you blinked!) This "persistence of vision" is what causes the two sides of the spinning flipper to merge into a single picture. The same effect explains how rolls of images flashing in front of our eyeballs at a steady speed turn into moving pictures, or movies.

An English doctor named John A. Paris popularized the flipper in 1824, calling it the thaumatrope. (The word means "turning marvel.") The toy carried the slogan "One good turn deserves another." Yet the discovery of the phenomenon that makes the flipper function is ancient. The Roman poet and philosopher Lucretius described persistence of vision in his work *On the Nature of Things* in 65 B.C.

Lucretius

For more flipper designs, visit www.potatochipscience.com.

LIDS

SOUND SPINNER

An Acoustical Device That Produces "Hole" Notes

FIELD: ACOUSTICS

CONCEPT: PITCH

MATERIALS

Pushpin

Chip lid
(included in kit)

6 feet of smooth string
(nylon twine works well)

Pencil

Hole puncher

METHOD

1 With a pushpin, poke two holes ½ inch apart at the center of the lid. Enlarge the holes slightly with a pencil so that the string will easily pass through them (see box on opposite page).

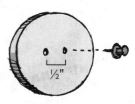

2 Feed one end of the string through both holes and tie the string's ends together to create a loop.

3 To wind up the spinner, position the lid in the middle of the string loop. Holding the loop loose on either side of the lid, whip your wrists forward, in a wheel-like motion, no fewer than 30 times.

4 To activate the spinner and create sound, gently pull your hands apart. When the string rewinds, allow the tension to bring your hands back together. (It should do half the work.)

Proper hand movement, when "playing" the sound spinner, resembles the technique of an accordionist.

In Colonial America, spinners were regularly crafted out of pierced copper coins.

... AND THIS

Rough edges cause friction. For the best results, keep the string holes smooth.

Too rough

Too small

Too big

Juusst right!

⑤ Once you've perfected the spin, use the hole puncher to make two holes across from each other at the edge of the lid and test your device again. Does the sound change? If so, how?

⑥ Now add two more holes at opposite sides of the lid's edge, and test the spinner once more. Does the sound change? If so, how?

MEANING

Sound spinners buzz with science. Their noises—the whirs, whizzes, and whines—are generated in the same way all sound gets made: through vibration. As the dancing chips on page 24 demonstrated, sound travels by the movement of air particles, which bounce against other air particles in a series of expanding waves. When those waves reach your ears, they bounce against your eardrums, which send signals to the brain that register as sound. The faster you spin

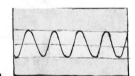

Low pitch (slow spin)

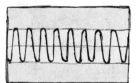

High pitch (fast spin)

your spinner, the higher the pitch it generates. Conversely, the slower you spin your spinner, the lower the pitch.

A wound-up sound spinner, much like a wound-up yo-yo, also demonstrates potential and kinetic energy (page 36). When the spinner is unwound (and potential energy turns into kinetic energy), the lid's momentum compels it to keep spinning. In fact, that momentum rewinds the string, setting things up for another noisy pull.

"COLOR" WHEEL

What's Black and White and Red all Over?

FIELD: NEUROSCIENCE
CONCEPT: OPTICAL ILLUSION

MATERIALS

"Color" wheel sticker
(included in kit)

Chip lid
(included in kit)

Map pin
(or longish straight pin)

Scissors

Drinking straw

Cork
(a rubber eraser
works, too)

METHOD

❶ Press a "color" wheel sticker into a chip lid.

❷ Poke the map pin through the center of the sticker and lid and enlarge the hole enough to allow the lid to spin without wobbling.

❸ With a pair of scissors, cut off a ¼-inch length of the straw and poke the pin through it. Then stick the pin into the cork.

❹ Spin the lid and stare at the sticker. Does it remain black and white or does it change color?

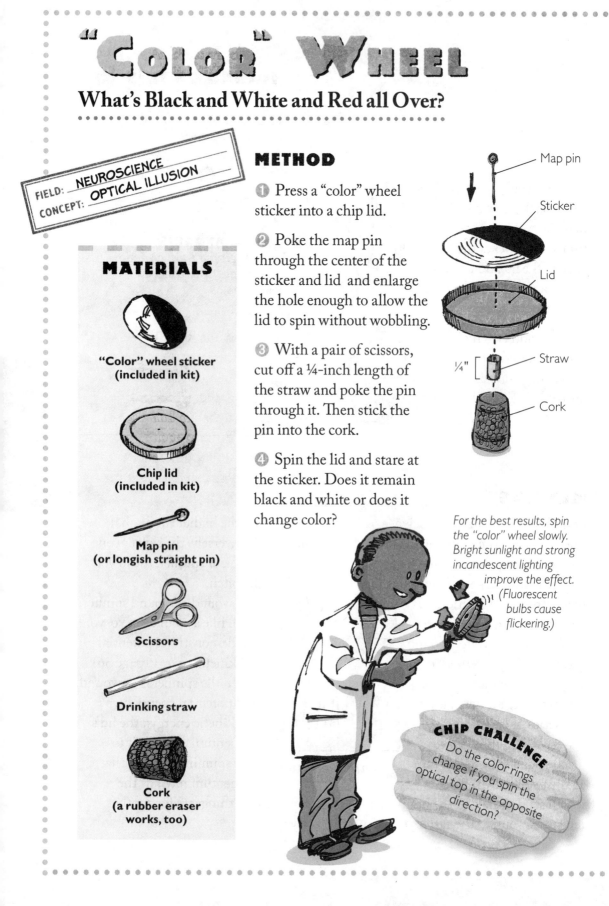

Map pin

Sticker

Lid

¼" Straw

Cork

For the best results, spin the "color" wheel slowly. Bright sunlight and strong incandescent lighting improve the effect. (Fluorescent bulbs cause flickering.)

CHIP CHALLENGE

Do the color rings change if you spin the optical top in the opposite direction?

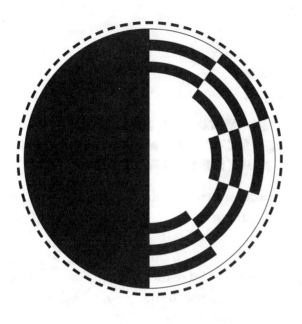

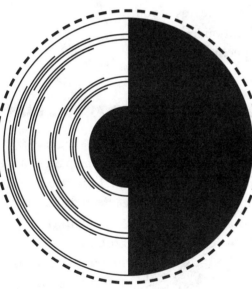

The "color" wheel first received worldwide attention in 1894, when C. E. Benham announced the invention of his "Artificial Spectrum Top."

MEANING

Why do the black-and-white lines of these optical wheels produce color when they're spun? No one knows for sure. The cause of this optical effect, known as the Subjective Color Illusion, is a mystery.

Some scientists argue that the effect is entirely optical (created only by the eyes). Others suggest it's produced by the eyes and the nervous system. Still other experts propose that the color switch is all in the brain. Although no one has proven why black-and-white patterns create the colors that they do, one very odd thing is certain: Different observers can see different colors emerging from the same spinning disc.

LIDS

MUNCH ON THIS

One of Benham's wheels was broadcast during the early days of television, and viewers perceived color images—on their black-and-white TVs.

SNACKTIVITY

When you find yourself in extreme conditions and furnished with only a few basics, you should never lose hope. Sometimes snack food is more than a snack. Sometimes it's survival.

HOW TO CURE A HEADACHE

Cut a potato into four pieces. Lie down in a dark room. Place two pieces of potato on your forehead, hold the other two pieces just above and in front of your ears (at your temples), and close your eyes. The pain should diminish. (There is no scientific explanation for this folk remedy—but herbalists swear it works like a charm.)

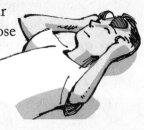

HOW TO STAY AFLOAT

If you're on a sinking chip ship and lack a life raft, don't despair. Simply duct-tape together 100 or so large, unopened bags of potato chips in two layers and position yourself on top.

HOW TO SIGNAL A RESCUE PLANE WHEN YOU'RE STRANDED

Fashion an unsinkable signal mirror by gluing the foil seal from the top of a chip tube to the lid of the tube. Then use the lid as a reflector to alert rescue teams of your location. (For more specific instructions, see page 38.)

HOW TO TREAT A MILD BURN

The *Old Farmer's Almanac* suggests this remedy for reducing the pain of a mild (first-degree) burn: Rub a freshly sliced potato over the affected area. The editors of the almanac explain that the starch neutralizes the burn and helps prevent scarring.

HOW TO KEEP YOUR EYE-GLASSES FROM FOGGING UP

If you're hiking through a tropical rain forest, you can stop your glasses from steaming up by lightly rubbing raw potato on the lenses.

SURVIVAL GUIDE

HOW TO STAY WARM AND DRY IN COLD, DAMP CONDITIONS

If you find yourself in a cold environment with nothing but the basics, don't despair. You can make yourself a dual-action rain poncho and emergency blanket out of 30 empty chip bags and some duct tape. The flexible, laminated chip bag material is entirely waterproof, and it keeps cold out by keeping warmth in. Chip bag emergency blankets form a reflective barrier that bounce 90 percent of the body heat back to the user. They also reduce sweat loss by slowing evaporation.

HOW TO TREAT MUSCLE SORENESS

Folk-remedy experts recommend tending to a sore foot or hand by soaking the affected limb in warm water that contains grated onion and potato. Professional bowlers have been known to reduce the minor swelling of injured fingers by cutting a hole in a potato and inserting the bruised digit.

HOW TO HEAT UP WATER WITHOUT FIRE IN THE DESERT

An empty chip bag serves as a fine solar thermal heater. If you place a bag of water in the sun, it will warm up quickly and retain heat nicely. (For added stability, roll the open end of the bag over on itself a couple of times.)

HOW TO ORIENT A MAP

Craft a compass out of a scrap of a chip bag, a potato chip lid, a magnet, a needle, and a little water. (For specific instructions, see page 5.)

HOW TO REMOVE RUST

Soak oxidized metal (fishhooks, for instance) in a solution of water and potato slices for two weeks. Rub away the nasty blemishes with a cloth. (Make sure you discard the used spud solution outside. Nothing creates a stench quite like 14-day-old potato water.)

SPUDS

Whether you make them snap (page 54),
crackle, or pop (page 58), a potato can take a prickin'
and keep on tickin' (page 50).

POTATO BATTERY

Getting a Charge Out of Spuds

FIELD: **ELECTRONICS**
CONCEPT: **ELECTROCHEMISTRY**

MATERIALS

2 copper electrodes
(included in kit)

2 zinc electrodes
(included in kit)

2 potatoes
(the bigger, the better)

Clock
(included in kit)

Wire connector
(included in kit)

Sound chip
(included in kit)

Chip tube

METHOD
To Make a Tater Clock

1 Jab one copper electrode and one zinc electrode into each potato. Keep the copper and zinc far away from each other.

2 Attach the clock leads (wires) to the electrodes in the following manner: Twist the red (positive) lead around the copper electrode of the first potato, and twist the black (negative) lead around the zinc electrode of the second potato.

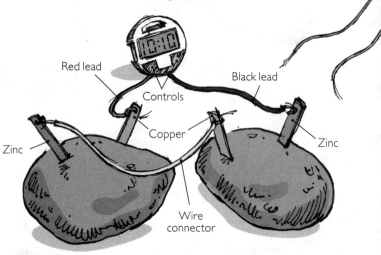

Red lead

Controls

Black lead

Copper

Zinc

Zinc

Wire connector

3 Complete the electrical circuit by twisting the exposed ends of the wire connector around the two unconnected electrodes.

4 Set the time using the two clock controls.

CHIP CHALLENGE

Try powering a clock off a couple of lemons. Can you find other fruits that will keep your clock ticking?

To Make a Potato-Powered Sound System

To run a sound system off a pair of potatoes, follow the method on the opposite page but swap the sound chip for the clock. To amplify the sound, stick the sound chip speaker to a metal-bottomed chip tube. Fiddle with the speaker to make the sound louder.

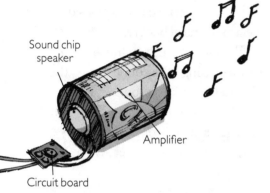

Sound chip speaker

Amplifier

Circuit board

Material Modifications

Spud clocks can be built out of all sorts of stuff. Here are some substitute substances:

• Zinc electrodes can be made from galvanized nails or shiny paper clips.

• Copper electrodes can be made from certain keys and coins. (Pennies minted before 1969, when the copper content was still very high, work especially well.)

• Wire connectors can be made from a rolled-up piece of aluminum foil.

Troubleshooting Your Tater: A Checklist

If your sound chip sounds weak you can wire more potatoes together for extra power. (Need additional electrodes? See Material Modifications to the left.)

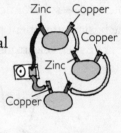

Zinc   Copper

Copper

Zinc

Copper

Still think you've got a dud spud? Not to worry. A few simple checks can often remedy the problem. Before you replace your batteries, make sure:

✔ Wires and electrodes are making contact (touching).

✔ Zinc and copper electrodes are spaced far apart.

✔ Each potato has one zinc and one copper electrode.

✔ Electrodes are clean. (Rust and oxidation can be removed with a little sandpaper or steel wool.)

Continued on next page

SPUDS

MEANING

Potatoes do not contain electricity. They are, however, full of electrolytes—chemicals that can be *turned into* electricity. With a few bits of metal, potatoes can be used as batteries—storehouses of energy that allow us to run digital clocks, tiny sound modules, and low-voltage LED lightbulbs.

Here's how a spud battery works: Poking copper and zinc strips of metal (electrodes) into a potato produces a series of chemical reactions that release electrons—subatomic particles that are the basic component of electricity.

Zinc sheds electrons more quickly than copper does. This imbalance triggers a steady movement of the electrons between the copper and zinc electrodes. When electrodes are looped in an alternating pattern (copper-zinc-copper-zinc) they create an electrical circuit. In effect, the metal probes and wire allow us to turn a chemical reaction into electrical energy. This is why a potato can be considered an electrochemical battery.

The more spuds you line up in a series—using wire connectors—the more electricity you can produce. Two potatoes provide just enough energy to power low-voltage electronic devices. Twenty can power a simple computer server.

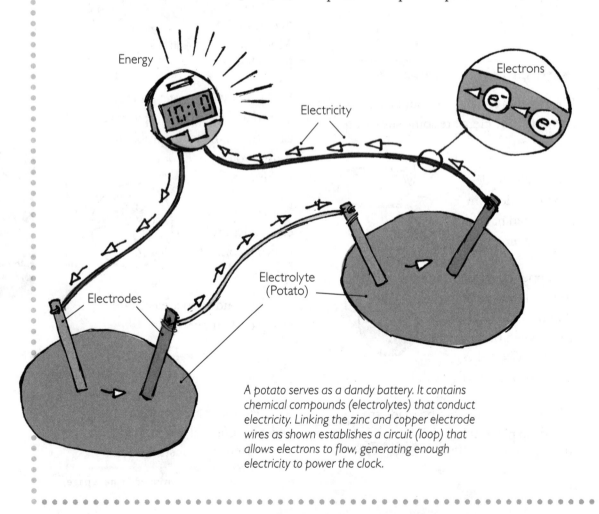

Energy

Electricity

Electrons

Electrolyte
(Potato)

Electrodes

A potato serves as a dandy battery. It contains chemical compounds (electrolytes) that conduct electricity. Linking the zinc and copper electrode wires as shown establishes a circuit (loop) that allows electrons to flow, generating enough electricity to power the clock.

Explaining Electricity (with the Greatest of EEEEEEs)

Electricity	A form of energy characterized by the flow of electrons. This flow can be generated in many ways, including (as in the case of the potato battery) chemical change.
Electrochemistry	The branch of science that deals with the production of electricity by chemical change.
Electrode	A terminal that conducts an electrical current into or out of potatoes and other fuel cells.
Electrolyte	A chemical compound that can conduct electrical current.
Electron	A negatively charged particle that spins around an atom. The flow of electrons produces electricity.
Energy	Usually defined as the ability to do work. Energy comes in many forms, including light, heat, sound, and motion. All energy is either kinetic or potential (page 36).

MUNCH ON THIS

The average baking potato stores about 1 volt of potential energy, which is why electrical devices that require no more than 2 volts of power can be run off a pair of large potatoes.

... AND THIS

Benjamin Franklin coined the word *battery* after a series of jars he used during electrochemical tests reminded him of a battery, or grouping, of cannons.

... AND THIS

You can connect potatoes in a series to increase voltage. One scientist built a spud battery to power a portable sound system. He needed 500 pounds of potatoes and a U-Haul truck to move the stereo around town. Talk about Eye-Tunes!

... AND THIS

Electrochemical batteries can also be made out of lemons, jars of vinegar, pots of wet dirt, soda pop, sports drinks, fruit juice, and other substances rich in electrolytes.

SPUDS

POTATO BENDER

An Absorbing Test That's a Snap to Perform

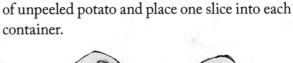

FIELD: MOLECULAR CHEMISTRY
CONCEPT: OSMOSIS

MATERIALS

Potato knife
(included in kit)

Potato

2 containers
(chip cans or small bowls)

2 cups water

3 tablespoons salt

METHOD

① Using a potato knife, cut two ¼-inch slices of unpeeled potato and place one slice into each container.

¼"

② Fill each container with a cup of water. Add the 3 tablespoons of salt to only one container and stir to dissolve.

Salt

③ Wait 30 minutes.

④ Remove the two slices and compare their textures. How much can you bend the potato soaked in the salt solution? How does its bendiness compare to the flex of the potato soaked in plain tap water? Does either slice snap when bent into a C?

CHIP CHALLENGE

Does a potato shrink or expand when placed in salt water? Can you come up with a test using pencil and paper to record a change in size? (Answer is at the bottom of the opposite page.)

MEANING

A potato slice soaked in tap water stays stiff (turgid). It breaks with a *snap!* when you try to bend it. On the other hand, a potato slice soaked in salted water, known as saline solution, turns rubbery (flaccid). Why?

Turgid

Flaccid

The reason is *osmosis*, the process by which water molecules move from an area of high concentration to an area of low concentration. This osmotic rebalancing act takes place because the cell walls of potatoes are semi-permeable. That means they permit water molecules to move in and out while keeping other molecules (such as salt) from passing through the cell walls.

If put into a salt solution, a potato will release water molecules to dilute the salt concentration surrounding it. Water from the spud makes the saline solution less salty.

If a potato slice is placed in everyday tap water, the potato draws in water and stiffens.

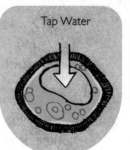

Tap Water

In a hypotonic environment, water passes into the potato cell.

Salt Water

In a hypertonic environment, water passes out of the potato cell.

SPUDS

Osmosis: A Mini-Glossary

In a **Hypertonic** solution, water *leaves* the cell by osmosis, causing it to become flaccid.

In a **Hypotonic** solution, water *enters* the cell by osmosis, causing it to swell and become turgid.

In an **Isotonic** solution, water does not move in or out of the cell.

SPUDDY BUDDY

How to Grow a Potato Pal

FIELD: BOTANY
CONCEPT: VEGETATIVE PROPAGATION

MATERIALS

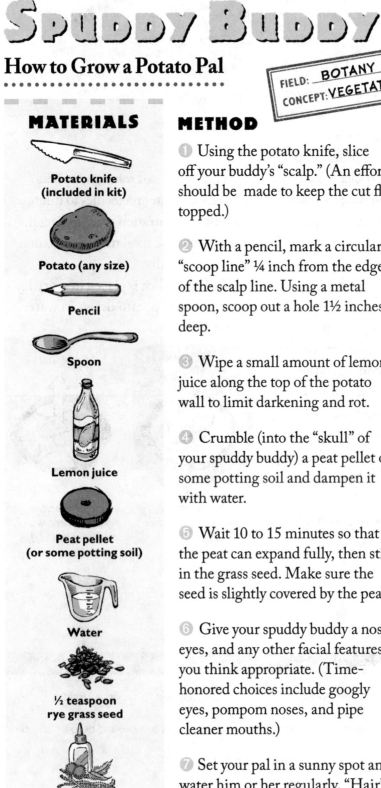

Potato knife (included in kit)

Potato (any size)

Pencil

Spoon

Lemon juice

Peat pellet (or some potting soil)

Water

½ teaspoon rye grass seed

Decorating supplies

METHOD

1 Using the potato knife, slice off your buddy's "scalp." (An effort should be made to keep the cut flat-topped.)

Scalp line

2 With a pencil, mark a circular "scoop line" ¼ inch from the edge of the scalp line. Using a metal spoon, scoop out a hole 1½ inches deep.

¼"

Circular scoop line

3 Wipe a small amount of lemon juice along the top of the potato wall to limit darkening and rot.

Rub lemon here

4 Crumble (into the "skull" of your spuddy buddy) a peat pellet or some potting soil and dampen it with water.

5 Wait 10 to 15 minutes so that the peat can expand fully, then stir in the grass seed. Make sure the seed is slightly covered by the peat.

6 Give your spuddy buddy a nose, eyes, and any other facial features you think appropriate. (Time-honored choices include googly eyes, pompom noses, and pipe cleaner mouths.)

7 Set your pal in a sunny spot and water him or her regularly. "Hair" can appear in as little as three days.

Answer: There are two cat eyes, two googly eyes, and lots of potato eyes.

Provide haircut as needed.

Day 1   Day 4   Day 7   Day 9   Day 10   Day 12

MEANING

All animals (from potato chip scientists to chipmunks) eat living things (whether plants or other animals) to fuel their bodies. Plants, however, are different. They obtain energy from sunlight. Each blade of grass that sprouts on the top of your spuddy buddy is a natural solar panel containing special cells that produce energy, in a process called photosynthesis.

The potato isn't a fruit and it isn't a root vegetable. It's a special kind of modified underground shoot known as a tuber.

All your potato pal needs to survive is water, nutrients from the peat moss, carbon dioxide in the air, and sunlight. While most plants can be grown only from seeds, potato pals, once they grow shoots, can be planted directly in the ground to generate more potatoes. The process by which plants grow from parent plants (instead of seeds) is called "vegetative propagation."

MUNCH ON THIS

Do you need potato seeds to grow potatoes? Absolutely not. Once the shoots of your potato pal grow a couple of inches, try planting him (or her) in a pot full of soil. Leaves should appear in about a week. Water the soil so that it's damp but not soaked. Place the pot in a warm, sunny location. When it's warm outside (and there's no risk of nighttime frost) you can transplant your potato pal in the ground, and watch your spuddy buddy turn into a spudmaker!

How many kinds of eyes appear in this picture? (Answer is on the opposite page.)

To see how your spud crop can power electrical appliances, turn to page 50.

SPUDS

FIELD: **PNEUMATICS**
CONCEPT: **BOYLE'S LAW**

PROPULSION PIPE
Our Mini Extermi–tater Is Highly *POP*ular

AN **ADULT** MUST CHIP IN

METHOD

TO LOAD

1 Cut a slice of potato ½ inch thick with the potato knife. (If you don't have a child-friendly potato knife, ask an adult assistant to take care of the slicing.)

⌐ ½"

2 Press one end of the pipe into the potato slice, then pull the pipe back up. A spud plug (chunk of potato) should remain inside the pipe.

First spud plug

3 Press the other end of the pipe into the potato slice, and again pull the pipe back up. You should now have a spud plug at each end of your pipe.

Second spud plug

TO LAUNCH

Using the eraser end of an unsharpened pencil, push one spud plug toward the other plug. Pressure will build between the two plugs until the top one goes flying. (It might take a few practice launches to perfect your technique.)

POP!

MATERIALS

| **Potato knife** (included in kit) | **Potato** | **Propulsion pipe** (included in kit) | **Pencil** (unsharpened) |

MEANING

A loaded propulsion pipe is packed with science. When you push two potato plugs closer and closer together inside the pipe, the volume of air shrinks and the pressure builds until the top plug is sent flying with a sonic *pop!*

The explosion confirms Boyle's Law, which states that volume and pressure are *inversely* proportional. This means the less there is of one, the more there is of the other.

The propulsion pipe also demonstrates the benefits of pneumatics—the use of pressurized gas (in this case, air)—to trigger mechanical motion. Nail guns, jackhammers, and the brakes on 18-wheel trucks all rely on pneumatic power.

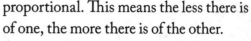

Robert Boyle

As the pencil pushes the two plugs closer and closer, the volume of air gets smaller and the pressure gets bigger.

● Keep your fingers away from the beveled edge of the propulsion pipe. It's pretty sharp.

● Never point a loaded pipe at a living creature.

● Wash out the pipe after every launch session.

● Pick up stray plugs. (Rotting ammo reeks.)

SPUDS

CHIP CHALLENGE

Try testing the range of apples, pears, and other firm fruits.

SHRUNKEN (POTATO) HEAD

AN ADULT MUST CHIP IN

FIELD: BIOLOGY
CONCEPT: DESICCATION

A New Wrinkle on an Ancient Recipe

MATERIALS

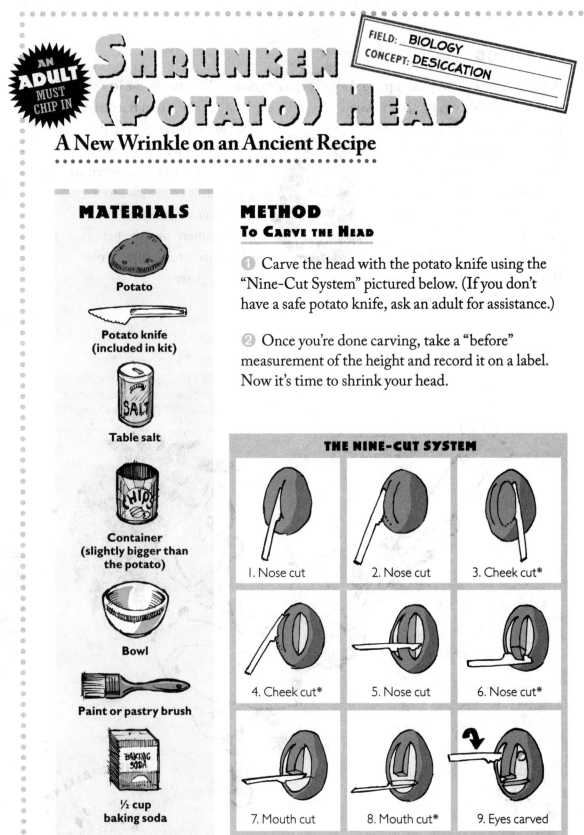

Potato

Potato knife
(included in kit)

SALT

Table salt

CHIPS

Container
(slightly bigger than
the potato)

Bowl

Paint or pastry brush

BAKING SODA

½ cup
baking soda

METHOD
TO CARVE THE HEAD

1 Carve the head with the potato knife using the "Nine-Cut System" pictured below. (If you don't have a safe potato knife, ask an adult for assistance.)

2 Once you're done carving, take a "before" measurement of the height and record it on a label. Now it's time to shrink your head.

THE NINE-CUT SYSTEM

1. Nose cut

2. Nose cut

3. Cheek cut*

4. Cheek cut*

5. Nose cut

6. Nose cut*

7. Mouth cut

8. Mouth cut*

9. Eyes carved

*Remove slice of potato

TO SHRINK THE HEAD

1 Place ½–1 inch of salt in the bottom of the container.

2 Center the head so that none of it touches the sides of the container. Then bury the head in salt. (Mix the baking soda into the salt for improved results.)

3 Once a day, empty the salt from the chip container into the bowl and, using the brush, remove any crust that forms around the head. Then, rebury the head in the container using the salt from the bowl.

4 If, after nine days, the head is still moist and spongy, ask an adult assistant to speed up the dehydration (drying) by placing your trophy in an oven preheated to 170 degrees Fahrenheit. Monitor the developments. The head should be nicely preserved in 3 to 4 hours, and begin to harden the following day.

5 Take an "after" measurement, label, and display.

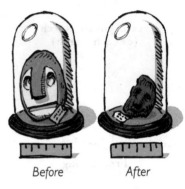

Before   *After*

MEANING

Technically speaking, your potato head isn't shrunken. And it isn't mummified either. To be strictly accurate, it's desiccated. That means the moisture has been sucked out with salt. This process is another example of osmosis (page 55). Removing moisture slows down the process of organic decay. In fact, properly desiccated potato heads will last for decades, if not centuries! There are other ways to preserve a spud head.

DRIED IN
SALT 9 DAYS
6.5 cm

You can use a dehydrator, mummify the head in baking soda, or, if you don't require massive shrinkage, air-dry your carved head in the sun. (The last method might require a month for the wrinkles to turn acceptably gross.)

Real shrunken heads come from the remote regions of Peru and Ecuador. These gruesome trophies, known as *tsantsa*, were the specialty of the Jivaro, a fierce Amazonian clan who, until about 100 years ago, regularly boiled their enemies' heads down to the size of baking potatoes.

SPUDS

SpudCrud

Is Potato Slime Liquid or Solid?

MATERIALS

**½ cup potato starch
(or corn starch)**

Freezer bag

½ cup water

**Food coloring
(optional)**

Large plate

METHOD

① Put the potato starch into a freezer bag.

② Pour about half of the water (¼ cup) into the freezer bag and zip it closed.

Adding optional food coloring is *highly* recommended.

③ Mush the contents together with your fingers.

④ Add more water, a little at a time, until the mixture is no longer crumbly. (You'll probably need less than the ½ cup water for this mixture. How much less water depends on how oozy you like your slime.)

⑤ Transfer the slime from the bag to a large plate. Is the mixture a liquid or a solid?

⑥ Now poke the slime and squish it with your fingers. Is it a liquid or a solid?

⑦ What happens when you stop poking the slime?

MEANING

Scientists generally define a solid as a material that keeps its shape and volume. A liquid, on the other hand, is defined as a material that maintains its volume but takes on the shape of its container. So what, then, is "spud crud?" That all depends.

The cellular structure of spud crud allows it to switch forms. Stabbing the surface of the slime with your finger will make it behave like a solid. Leaving the slime alone allows it to return to a liquid state.

Substances that display this ability to change form are known as non-Newtonian fluids. That's because Sir Isaac Newton argued—incorrectly—

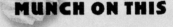

Isaac Newton

that all liquids act the same whether held or observed. (Sorry, Isaac.)

Spud crud oozes with other insights into molecular chemistry as well. Structurally speaking, spud crud is considered a polymer. Polymers are very large molecules formed by repeated patterns of chemicals strung together. Some polymer chains are artificially produced, while others occur naturally. Bone is a polymer. So is blood, rubber, cotton, and Jell-O. Plastic potato chip tubes, it must be noted, are also made from polymers. And so are potato chip bags. (Turn the page to find out more.)

- Keep slime *away* from pets and small children.

- *Handle slime with care. It makes a mess.*

- *Refrigerate slime when not in use to avoid slime stench.*

- *Never pour slime down a drain, because it will clog the pipes. Discard it in the trash instead.*

MUNCH ON THIS

Bulletproof armor is filled with a special non-Newtonian fluid (not spud crud) that is soft and flexible— until it's hit by a bullet. Then it turns solid.

SPUDS

THE LIFE CYCLE OF

The pleasure of eating a potato chip is simple. Delivering it to your lips is not. Every bag of chips contains a good deal more within its foil-lined interior than a deep-fried, thin-sliced, salted snack food. It's also cram-packed with science!

GROWING THE POTATOES

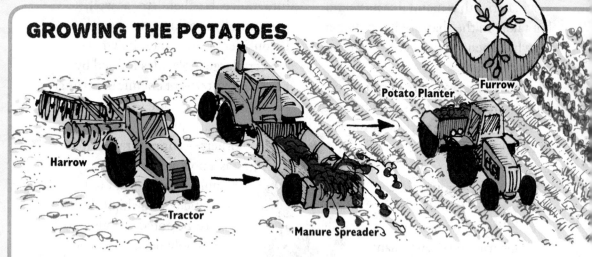

Harrow

Tractor

Manure Spreader

Potato Planter

Furrow

PREPARATION & PLANTING: After the first frost, the farmer digs up and levels the potato field using a tractor and harrow, and then fertilizes the soil with a manure spreader. Once that's done, a potato planter drops seed potatoes into the soil, which is shaped into mounds called furrows. Seed potatoes, planted whole or cut into pieces, generally contain one to three active eyes.

HARVEST & STORAGE: Most potatoes are ready to harvest three to four months after they are planted.

MAKING THE BAGS

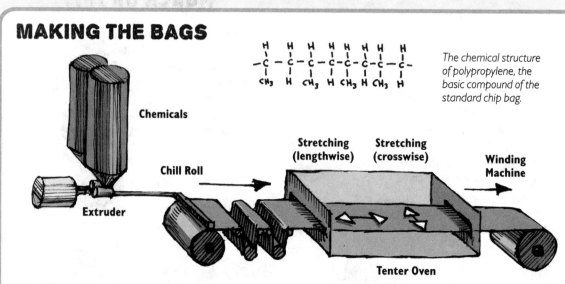

Chemicals

Chill Roll

Stretching (lengthwise)

Stretching (crosswise)

Winding Machine

Extruder

Tenter Oven

The chemical structure of polypropylene, the basic compound of the standard chip bag.

FILM MAKING: The production of potato chip bags starts by combining various chemical compounds containing carbon and hydrogen into a resin called a polymer. That polymer (polypropylene) gets melted and stretched in a tenter oven to produce rolls of super-thin, multilayered material designed to keep your chips safe and fresh.

A POTATO CHIP

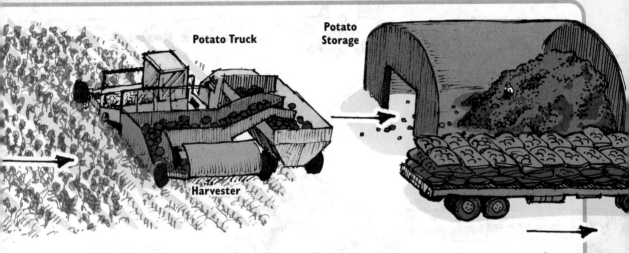

Potato Truck

Potato Storage

Harvester

State-of-the art equipment can unearth and load 16 tons of potatoes in 3½ minutes. Potatoes often remain on the farm for months. During that time, farmers inspect and sort the crop by size, then store it in warehouses that are cool, moist, and dark.

Improper storage can turn potatoes green, a sure sign that they've picked up a natural toxin (poison) called solanine. Large facilities can store more than 20,000 tons of spuds.

Anatomy of a Chip Bag: Pick up an old chip bag and look at it. Closely. What you're holding in your hand is a complex feat of structural engineering: a seven-layer cake of plastics and metals cooked up to protect a very fragile cargo.

Working from the outside of the bag to the inside, here's how those seven layers stack up:

1. An outer transparent heat-sealable film.

2. An outer plastic core.

3. A print skin. (Chips bags get printed backwards so that the words appear correctly when viewed by snackers.)

4. A stiffening layer that gives a bag strength, stability, and form.

5. A metallized film—the so-called foil—composed of super-thin aluminum about 60 atoms thick. The metal layer is applied to block out light, the enemy of crisp potato chips.

6. An inner plastic core.

7. An inner transparent heat-sealable film, which gets melted with the outer film to form a seal.

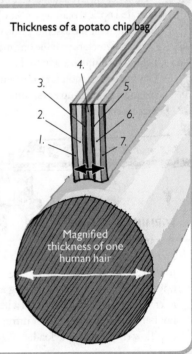

Thickness of a potato chip bag

Magnified thickness of one human hair

Continued on next page

MAKING THE CHIPS

LOADING DOCK: Potato chips require potatoes, oil, and salt. Trucks or freight cars bring these items, plus the packaging they're sold in, to the factory loading dock.

Inspection Station

INSPECTION: As soon as spuds arrive at the chip factory, they're inspected for size, color, taste, moisture, and sugar content. (Sweetness and moisture affect the cooking process.) Approved spuds are sent to a storage facility. Rejected spuds are usually rendered into potato starch and pig feed.

Continuous Fryer

Metal Detector

Optical Scanner

Seasoning Tumbler

SEASONING & SURVEYING: Cooked chips move on to the cooling conveyor and from there to the seasoning tumbler, where various salts and flavors are applied.

Either eagle-eyed inspectors or computer-operated optical scanners isolate "reject" chips and blow them off the assembly line with a puff of air.

Metal detectors provide one more safety check before the chips are sent to the bagging station. (It's not uncommon for the assembly-line safety magnets to pluck up keys, coins, and broken bits of machinery.)

FRYING: Potato chips are generally fried either in kettles or in troughs. Kettle chips, which tend to be thicker and crunchier, are immersed in hot oil for about 7 minutes—twice the time needed to fry a potato chip in the trough of a continuous fryer.

To verify that the seams of bags are air-tight, samples are regularly placed in burst chambers or submerged in water.

Bagging Machine

Sampling Station

Boxing Station

FORMING & FILLING: Bag material, which arrives at the chip factory in rolls, is formed into an open bag. Chips drop into the bag and seconds later the top is melted shut with a heat seal—but not before a shot of gas (air or nitrogen) is added to pillow the delicate payload.

CHECKING FOR QUALITY: Chips, like most foods, undergo one last inspection before they leave the factory. Samples are tested for appearance, texture, and taste. Chip color is compared to charts. Bag seals are tested, too.

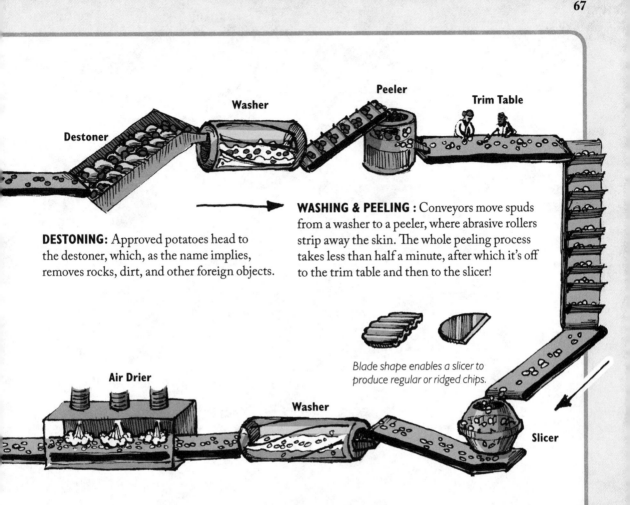

Destoner **Washer** **Peeler** **Trim Table**

DESTONING: Approved potatoes head to the destoner, which, as the name implies, removes rocks, dirt, and other foreign objects.

WASHING & PEELING : Conveyors move spuds from a washer to a peeler, where abrasive rollers strip away the skin. The whole peeling process takes less than half a minute, after which it's off to the trim table and then to the slicer!

Blade shape enables a slicer to produce regular or ridged chips.

Air Drier **Washer** **Slicer**

DRYING: Conveyors pass chips under the jets of an air dryer to remove excess water. Then it's time for the fryer.

SLICING: When a raw spud drops into the slicer, it's pressed against the inner wall of the cutting drum, much like a human body is pressed against the cylindrical wall of an amusement park Tilt-A-Whirl. There's one big difference, though. This drum wall is fitted with razor-sharp blades! As the spud passes across the knives, it's cut into slices $\frac{1}{20}$ of an inch thick. A slicer can work through 5,000 pounds of potatoes an hour.

DISTRIBUTION & CONSUMPTION (AKA SNACKING): Bagged chips are boxed and dispatched in trucks, which were once equipped with special shock absorbers to reduce breakage during shipping. Chip makers avoid transport by airplane because dramatic changes in atmospheric pressure can cause an entire shipment of chip bags to explode.

Snack Enthusiast

WASTE DISPOSAL: Potato chip packaging has three possible destinations: landfill, the recycling plant (which can process tubes and lids), or the laboratory of the potato chip scientist. We hope this book will prompt you to choose option three.

Landfill

Recycling Plant

Laboratory

TUBES

What makes birds chipper,
launches confetti, gobbles, clucks, kisses,
and does a fine job teaching physics?

(Hint: The answer is staring you in the face!)

BIRD FEEDER

From Chips to Chirps in Six Simple Steps

FIELD: ORNITHOLOGY
CONCEPT: BIRD-WATCHING

MATERIALS

Clean chip tube with lid
(only plastic ones will work)

Paper towels

Marker

Craft knife

Screw eye

Stubby pencil
(about 5 inches long)

Ribbon

Wild birdseed

METHOD

1 Wash out and dry the tube with a paper towel. (Salt can harm birds.)

2 With the marker, draw a door (an upside-down U) near the bottom of the tube and ask an adult to cut along the marks with the craft knife.

3 Use the screw eye to twist a starter hole below the door, and force the stubby pencil into the hole so that it sticks out about 2 inches.

2"

4 Twist the screw eye into the center of the lid, and tie the ribbon to the eye loop.

5 Lay newspaper under the feeder to catch stray seeds, then fill it with the birdseed (no more than three-quarters full). Press the door slightly inward, into the tube.

— Fill line

Newspaper contains spills.

6 Replace the lid—adding a little tape if necessary—and hang your feeder outside using the ribbon.

Keep seed dry. Moldy grain is bad for birds.

MEANING

What kind of bird will be drawn to your feeder? That's hard to say. It depends on where you live, the season and what's in your seed mix. Different species of birds inhabit different regions of the country. Plus, it's important to remember, birds' diets vary widely. Blue jays love corn. Nuthatches go nuts for peanuts.

Feeder design will also affect the guest list. Tray feeders placed on or near the ground attract juncos, doves, and sparrows. Suet feeders bring in starlings and woodpeckers. Hummingbirds are suckers (literally!) for sugared water.

It will probably take a few days (maybe even a week) for your feeder to attract feathered visitors, but once they do arrive, they'll keep coming back.

Serious bird-watchers always keep three things handy: a journal for recording observations, a guidebook to help with identification, and binoculars.

MUNCH ON THIS

Cardinal   Chickadee   Finch   Titmouse

When filled with mixed seed, your feeder should keep cardinals, chickadees, finches, and titmice very, *very* chipper!

FIELD: **PHYSICS**
CONCEPT: **ACOUSTICS**

AN ADULT MUST CHIP IN

CHIP-TUBE GOBBLER

Building This Birdcall Is a Real Hoot!

MATERIALS

Hammer

Thin nail

Short chip tube (metal-bottomed ones work best)

Paper clip

18 inches of string

Small piece of damp sponge

Decorating supplies (googly eyes, pom-poms and feathers are highly recommended)

METHOD

1. Use the hammer and nail to poke a center hole in the bottom of the tube.

2. Tie one end of the string to the paper clip and feed the loose end through the center hole so that the paper clip remains on the outside of the tube.

3. To make your gobbler gobble, hold the tube with the bottom side up (paper clip on top), and drag a small piece of damp sponge down the string in short, jerky movements. (Place your finger on the paper clip and you'll produce a clearer sound.)

4. Once you've mastered the technical aspects of the birdcall, give the device a beak and wings (use the template on the opposite page), then decorate it.

CHIP CHALLENGE

Can you modify your birdcalling technique to produce a cluck? A hoot? A growl?

gobble!
gobble!

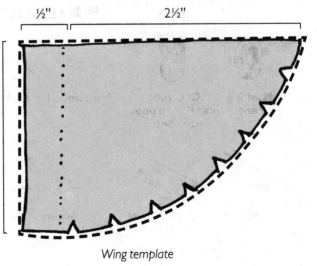

Don't use coarse twine on your gobbler. Smooth string works better.

MEANING

Dragging a damp sponge down a string sends invisible air molecules outward in waves. Those air molecules bump into other ones, which in turn bump into still more. When these waves reach your eardrum, a signal is sent to your brain and you hear sound. Sound waves come in various shapes. Low-pitched sounds, like the strum of a bass guitar, make broad vibrations. High-pitched sounds, like the shrill toot of a pennywhistle, produce more closely spaced, shorter waves (page 43). The chip tube serves as an amplifier—meaning that it increases the sound. The amount of pressure you put on the string and how you squeeze the sponge, plus the speed, length, and smoothness of your downward tug, will change the sound your gobbler makes.

To complete the gobbler, photocopy the beak template once and the wing template twice, or trace a beak and two wings on a sheet of paper. Cut along the outside dashes and fold along the inside dotted lines. Then glue the body parts to the tube.

½" 2½"

1¼"

1¼"

2"

Beak template

Wing template

TUBES

FIELD: **PHYSICS**
CONCEPT: **BALLISTICS**

The word cannon *comes from the Latin* canna, *which means* tube.

CONFETTI CAN-NON

A Lighthearted Launcher of Ultralight Missiles

METHOD

TO PREPARE THE CAN-NON BARREL

① Wrap a piece of tape around the chip tube ½ inch from the bottom and, with the pushpin, poke holes every ¼ inch around the edge of the tape line.

② Next, cut between the holes with the potato knife to remove the bottom. (If a potato knife is not available, ask an adult to do the cutting.)

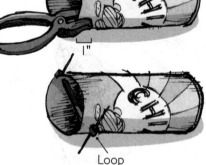

③ With the hole puncher, punch two holes opposite each other about 1 inch up from the bottom of the chip tube.

④ Feed the rubber band through the holes, leaving a loop hanging out on each side.

⑤ Insert a toothpick through each loop and pull on the rubber band to tighten it.

Loop

MATERIALS

Masking tape   Chip tube (cardboard ones work best)   Pushpin   Potato knife (included in kit)   Hole puncher   Large rubber band

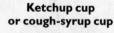

2 toothpicks   Paper towel tube   Scissors   Ketchup cup or cough-syrup cup   Chip-bag confetti

TO PREPARE THE LAUNCH TUBE

① Punch two holes opposite each other about 1 inch in from one end of the paper towel tube. Then cut a slit from that end of the tube to each hole.

1"

② Press a ketchup cup into the other end of the paper towel tube. (Tape it in place if it's loose.)

③ Feed the hole-punched end of the paper towel tube down through the top of the chip tube until the slits catch the rubber band strung across the chip tube's bottom.

④ Fill the cup with chip-bag confetti (page viii). Reach into the barrel, pull back on the rubber band, aim away from anything that will whimper, bite, cry, tattle, yell, break, or topple ... and let fly!

Rubber bands lose their stretch over time. Use only fresh ones when building your can-non.

MUNCH ON THIS

Besides confetti, the *can*-non is able to launch all sorts of missiles: potato starch packing "chips," Ping-Pong balls, bits of sponge, feathers, earplugs ...

MEANING

To launch confetti (or anything else) you need enough *oomph!* (force) to counteract the downward pull of gravity. The *oomph!* of this *can*-non comes from the potential energy stored in the stretched rubber band. When you release the band, potential energy is transformed into kinetic energy. The result: The missiles (also called projectiles) shoot through the air in a curved path known as a trajectory. Unless they manage to reach outer space, all projectiles—from bullets to baseballs—end their travels back on Earth, thanks to gravity's pull.

TUBES

KISSING TUBES
A Low-Pressure Challenge

FIELD: **PHYSICS**
CONCEPT: **BERNOULLI'S PRINCIPLE**

MATERIALS

12 drinking straws

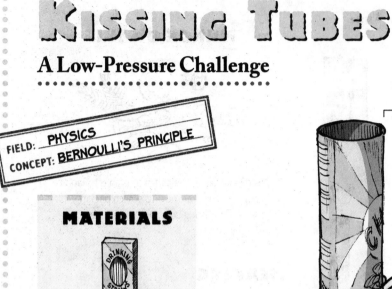

**1 pair of chip tubes
(cardboard, any height)**

METHOD

1 Line up the straws on a flat surface, spacing them ½ inch apart.

2 Place the chip tubes upright on the straws, leaving 1 inch of space between them.

3 Now here's the low-pressure challenge: Can you make the two tubes touch with a puff of breath? (Answer is at the bottom of the opposite page.)

MEANING

It's not magic that makes the chip tubes "kiss" when you blow between them. It's physics. Fast-moving air produces lower pressure than does slow-moving air. Your breath decreases the air pressure between the tubes. That causes the higher pressure elsewhere to push the tubes together, a phenomenon first explained by Daniel Bernoulli (page 37).

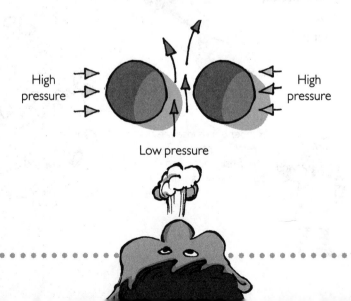

High pressure   High pressure

Low pressure

AN **ADULT** MUST CHIP IN

WALKIE-TALKIE

FIELD: __ACOUSTICS__
CONCEPT: __TELEPHONY__

Finally! A Communications Network That *Isn't* Wireless

MATERIALS

| Hammer | Thin nail | 2 short chip tubes (metal-bottomed ones work best) | Thin wire (30 feet) | 2 paper clips |

METHOD

❶ Use the hammer and nail to poke a center hole in the bottom of each tube.

❷ Insert one end of the wire through the bottom of the first tube and twist that end around a paper clip.

❸ Insert the other wire end through the bottom of the second can. Again, secure the wire end with a paper clip.

❹ Have two researchers, each holding a tube, move away from each other until the wire between them is stretched taut. (This is the "walkie" part.)

❺ Have the two researchers alternate speaking and listening. (This is the "talkie" part.)

MEANING

These walkie-talkies reveal the pre-digital roots of telephone technology (telephony)—when phone lines allowed communication only between two "hard-wired" handsets. Yet, as simple as these voice devices might be, they work better than most cell phones. That's because sound vibrations travel more efficiently through a solid such as wire than they do through air. The tubes of this voice-transmission system amplify the sound that travels along the wire.

TUBES

Answer: To make the tubes touch, position your lips in front of the tubes and blow between them.

WINDMILL

These Construction Plans Are a Breeze

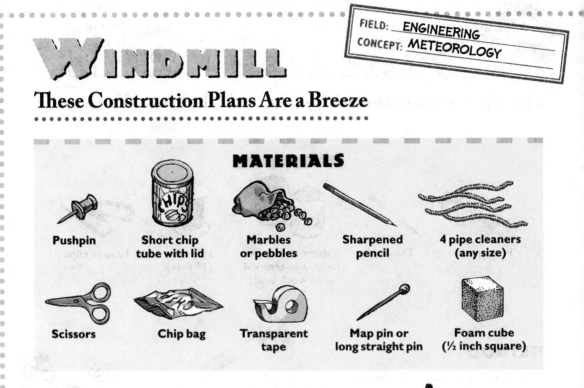

MATERIALS

Pushpin | Short chip tube with lid | Marbles or pebbles | Sharpened pencil | 4 pipe cleaners (any size)

Scissors | Chip bag | Transparent tape | Map pin or long straight pin | Foam cube (½ inch square)

METHOD
To Make the Base

1. Pierce the center of the chip lid with the pushpin.

2. Fill the tube with the marbles for stability and replace the lid.

3. Insert the point of the pencil through the hole in the lid and push it straight down so that the pencil touches the bottom of the tube and stands upright.

To Make the Blades

1. Bend each of the 4 pipe cleaners as shown.

2. With a pair of scissors, cut 4 rectangles out of the chip bag, each 3 x 2 inches.

3. Tape 1 pipe cleaner to each rectangle. These are the blades of your windmill.

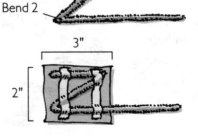

Bend 1

Bend 2

3"

2"

TO ATTACH THE BLADES

1 Pin the foam cube to the top of the pencil with the map pin. The cube should spin easily. (If it doesn't, wiggle the map pin to enlarge the hole a little.)

2 Push a pipe cleaner into each side of the cube.

3 Test the spin by blowing across the blades. Your windmill should turn easily.

4 Place your windmill outside on a windy day.

Foam cube

MEANING

Windmills have been around for more than 1,000 years and were originally used to grind grains and pump water. Modern-day wind machines are more often used to convert wind power into electrical energy.

What produces wind? Simply put, differences in air pressure. Here's what happens: The sun heats the surface of the Earth. The Earth releases that heat and warms the air above it. As warm air rises, cold air funnels down to take its place.

This movement between warm (low-pressure) air and cold (high-pressure) air, plus the rotation of the Earth, generates wind. Big differences between currents of warm and cold air create strong winds. The lower the pressure, the stronger the wind.

Anemometer   Weather vane   Wind turbine

Wind Machines: A User's Guide

Anemometer	Indicates strength of wind
Weather vane	Indicates direction of wind
Wind turbine	Generates energy from wind

TUBES

POTATO CHIP SCIENCE
How We Cooked It Up

I'll confess. Until a few years ago, I never thought much about potato chips. (Heck, I rarely even ate them.) The topic only grabbed my attention after my son, Max, then nine years old, challenged me to "forget history" and write about something *interesting*.

"Like what?" I asked, tossing the dare right back.

"Like the Red Sox," he teased, knowing I grew up rooting for a more talented team based some 200 miles south of Fenway Park. "*Next!*"

"What about potato chips?" he asked.

I balked. But initial hesitation slowly gave way to mild interest. And mild interest turned into fascination. That fascination eventually evolved into a full-fledged scientific partnership that offered a father and son deep and lasting satisfaction. Simply put, the chip off the old block turned the old block into a chip fiend by proposing a project we could both sink our teeth into.

The chip

The earliest public expression of this carb-fueled collaboration appeared in *Leon and the Spitting Image*, a children's book whose hero, like Max, loves potato chips. That novel led to a sequel, *Leon and the Champion Chip*, which celebrated the work of Franklin Sparks, a science teacher who cooks up a fifth-grade curriculum devoted to thin-sliced, deep-fried tubers. For Franklin Sparks's innovative lab work to ring true, Max and I were compelled to do some experimenting and testing on his behalf. Our scientific research began at the local Stop & Shop, where a $10 investment enabled us to study the relative acidity of chips. Next we tested the exothermic properties of chips, meaning we set them on fire. The residue of those combustion tests—black, sooty powder—ended up in the detective kit on page 21.

That was not the only by-product of our early investigations. We also ended up with a ton of empty bags, which we tried to turn into balloons. Those efforts failed, but the next invention—the bag blaster on page 2—proved more successful.

Soon we were broadening our inquiries beyond chips and bags by studying the scientific properties of the potato itself. That research led to the Extermi-tater—a 5-foot-long spud launcher powered by a hair-care product called Aqua Net. I tried to explain to Max's mother, who's from Paris, that the apparatus deepened Max's appreciation of Boyle's Law. "*Bien sur,*" she said, which is French for "Yeah, right."

After we rocketed a russet across the length of two baseball fields,

The old block

Françoise confiscated the hair spray. Undeterred, we reduced the scale of the launcher and made it less dangerous. The result: the pocket-sized propulsion pipe that accompanies this kit.

Not all spud work focused on physics. Electrochemistry also figured in our investigations. We pieced together a serviceable—if somewhat dorky—potato clock based on directions pulled off the Web. The aluminum foil didn't work all that well, nor did the pennies, so we modified the plans. (The results appear on page 50.)

And as often seems to happen in our household, research triggered a visit to eBay. Max was authorized to buy any and all chip-related research material, as long as it cost under $15 (including shipping). Thus the arrival of the "Valentine"—the rare heart-shaped chip showcased in the Chipatorium (page 30, fig. No. 2)—and, rarer still, "Chipper," a naturally formed smiley face that emerged from a bag of Miss Vickie's chips back in 1972.

Fast-forward to the present. I now seek out chip litter like a crazed prospector. Everywhere I go—at meetings, during Little League games, while walking our dog—my eyes rake the streets for empty bags, lids, and tubes.

And for proof of just how far it's possible to take the study of potato chips, consider the following:

I had just finished touring a potato processing plant during a research trip to Boise, Idaho, when I spotted Chipper—

Potato Chip Science can be a blast.

our Chipper—on TV. The same wrinkly face, the same jaunty smirk. That's impossible! I told myself. As it turned out, the smiling snack on the screen was Chipper's unacknowledged twin, cared for by a famous L.A. talk show host.

When I told Max, he insisted we take photos and send them to the show's producer. He wanted to arrange a rendezvous between the identical twins. We're still waiting for a response. And though

Our Chipper

it's embarrassing to admit, my son and I actually felt disappointment that the two chips—this is nuts, I know!—failed to meet.

So we took comfort the best way we knew. The same way billions of people all over the world take comfort. We reached for a bag of potato chips—a bag we quickly emptied and then refilled with all sorts of science.

—Allen Kurzweil

GLOSSARY

Acoustics: the study of sound

Aerodynamics: the study of the effect of the motion of gases (which include air) on objects

Anthropology: the study of humankind, especially human culture or human development

Ballast: a heavy material used to stabilize a ship or airship

Ballistics: the study of the movements and forces involved in the propulsion of objects through the air

Bernoulli's Principle: the physical law that states that the speed of a fluid and pressure in a fluid are inversely proportional (as the speed increases, the pressure decreases, and vice versa)

Biodegradable: able to decay as a result of the action of bacteria

Biology: the branch of science dealing with the study of living organisms, including their structure, function, growth, origin, evolution, and distribution

Bird-watching: the observation of birds in their natural habitats

Botany: the study of plants

Boyle's Law: the principle that the volume of a confined gas at constant temperature varies inversely with its pressure

Buoyancy: the tendency of a liquid or gas to cause less dense objects to float

Calorie: the amount of energy (heat) required to raise the temperature of 1 gram of water 1 degree Celsius

Chemistry: the branch of science dealing with the structure, composition, properties, and reactive characteristics of substances

Conservation of energy: the principle that states that the amount of energy in an isolated system remains the same, even though the form of energy may change

Desiccation: the process of extreme drying, often by means of salt

Digestion: the breaking down of food into a form that can be absorbed by a living body and used as energy

Electrochemistry: the branch of chemistry that studies chemical change associated with electrons and electricity

Electronics: the science and technology of electronic phenomena

Electrostatics: the branch of physics dealing with electric charges at rest (static electricity)

Engineering: the practical application of science to commerce or industry

Environmental science: the study of interactions among physical, chemical, and biological components of the environment

Food energy: the amount of energy in food available through digestion

Forensics: those branches of science and medicine that have a specifically legal purpose

Hydrodynamics: the branch of fluid dynamics dealing with liquids

Magnetism: the phenomenon by which materials exert an attractive or repulsive force on other materials

Meteorology: the study of Earth's atmosphere, especially its patterns of climate and weather

Microbiology: the study of microscopic organisms

Molecular chemistry: the study of how molecules (small particles) interact

Navigation: the art or science of plotting or directing the course of a ship, aircraft, or other vehicle

Neuroscience: the study of those aspects of the nervous system controlled by the brain

Non-Newtonian fluid: a fluid whose viscosity is variable based on applied stress

Nutrition: the act or process by which plants and animals nourish themselves, as well as the study of food and nourishment

Optical illusion: a false visual perception that tricks the brain

Optics: the study of light and vision

Ornithology: the branch of biology dealing with the study of birds

Osmosis: the flow of a substance through a semi-permeable membrane from a more concentrated solution to a less concentrated one

Persistence of vision: the phenomenon by which the brain holds on to an image for a longer time than it is displayed

Photosynthesis: the process by which plants (and some bacteria) transform sunlight into chemical energy

Physics: the branch of science dealing with matter, energy, force, and motion

Physiology: the biological study of the way organisms function

Pitch: the level of a sound in the scale, defined by its frequency

Polypropylene: a chemical resin used in many consumer products, including potato chip bags

Polymer: a large molecular substance, such as plastic, composed of repeating structural units

Pneumatics: the branch of physics dealing with the mechanical properties of air and other gases

Propulsion: the process by which an object is moved forward

Reflection: the process by which light, sound, or heat "bounces off" a substance

Rocketry: the design, construction, flying, and technology of rockets

Solanine: A toxin (poison) found naturally in certain plants, including green potatoes

Sound wave: the audible pressure caused by a disturbance in water or air and carried forward in waves

Surface tension: the property of liquids that gives their surfaces a "skin"

Telephony: the transmission of sound between different stations, especially by radio, telephone, or potato chip tubes

Vegetative propagation: a form of asexual reproduction by which new plants are generated from parent plants without seeds or spores

ACKNOWLEDGMENTS

As the cover warns, this book is unusually high in saturated facts. Nationally recognized experts from every branch of Potato Chip Science provided many of those facts. Kids provided even more. The names listed below may appear in tiny type, but the contributions of each and every bag-blasting, chip-burning, lid-launching, spud-snapping, tube-testing, pencil-wielding helpmate has been huge—and greatly appreciated.

BAGS Harley Frank, of Admiral Packaging ❦ Terry Baker and Len Limongelli ("It's pronounced 'Len Lemon Jelly!'") at C-P Converters ❦ Tom Bryce, the "film mogul" of Bryce Corporation, and his flexible sales rep Dan Conley ❦ Leighton Derr of AET Films, who detailed the structural properties of laminated bags ❦ Debra Jacobson at the Illinois Waste Management and Research Center, for her analysis of the impact of flexible packaging on landfill ❦ And Printpak executives Carole Anderson and Kimberly N. Carter, who provided a wonderful rap on wrapping **CHIPS** Jim Green, the Kettle Foods Chip Ambassador, whose assistance was exceeded only by that of his colleague Carolyn Richards, the company's sweet and savory flavor architect ❦ Al Greer, Jr., the undisputed übertuber of George Greer Foods ❦ Jane Rice of Utz Quality Foods ❦ Melissa Shearer of Shearer's Foods ❦ Leon Stoltz, Potatofinger's hands-on CEO, who kept our basement stocked with "research material" ❦ Daryl Thomas of Herr Foods, Inc. ❦ And to potato chip pastry chefs Lynn Williams of Seven Stars Bakery and Marcia Bernard, Librarian of the Shutesbury Elementary School, for turning their kitchens into science labs. (Their recipes can be found at www.potatochipscience.com.) **LIDS** Lisa Copes and Cheryl Johnson at Tech II Inc., who sent us a vast and colorful assortment of chip lids ❦ And tennis pro Larry Sack, who also served up 3-inch overcaps **SPUDS** Hasbro CEO Alan Hassenfeld, patriarch of (among other things) the Mr. Potato Head franchise, and his colleague Franklin Labarbara ❦ Ben Kudwa, Executive Director, Michigan Potato Industry Commission ❦ Daphne Pulver, who guided us through the largest potato-processing plant on the planet ❦ The folks at Simplot, who allowed a very happy potato chip scientist to ascend their potato mountain **TUBES** Florence and Lou Goeringer, the mom and pop of the Bertels Can Corporation, for their insight into old-fashioned potato chip tins ❦ Professor Donal O'Shea, the Elizabeth T. Kennan Professor of Mathematics at Mount Holyoke College, for his algebraic analyses **EDITORIAL** The chipper team at Workman, who made this kit possible: Savannah Ashour, Susan Bolotin, Andrea Fleck, Beth Levy, Nate Lifton, Randy Lotowycz, David Matt, Rae Ann Spitzenberger, Janet Vicario, Walter Weintz for the ornithological insight that transformed the chip-can clucker into a chip-can gobbler, Doug Wolff, and Peter Workman ❦ Literary agents Liz Darhansoff and Chuck Verrill ❦ Artist Bret Bertholf ❦ And Virginia Duncan and Susan Katz at HarperCollins **TECHNICAL** The guys at AGC Sound, for providing us with the low-voltage sound modules that simulated a variety of very beautiful and very rude noises ❦ Aero Rubber Company, for documenting the life expectancy of rubber bands ❦ Christine M. Cunningham at the Museum of Science, in Boston, MA ❦ Aseem Das, who introduced us to the environmental virtues of the potato knife ❦ Steve Johnson and Debra Novello of Urschel Laboratories, who documented cutting-edge breakthroughs in the blade design of chipping machines ❦ Charles Kingdon ❦ David Lewis, whose extraordinary chip bag glider design can be downloaded at www.potatochipscience.com ❦ Ken Lonngren at www.infolddesign.com ❦ Professor Steven Lubar, the unflappable and thoughtful director of the John Nicholas Brown Center for the Study of American Civilization ❦ Char Dailey ❦ Charles Melcher ❦ Professor Rosana Moreira of Texas A&M University, for the CAT scan of a chip bag ❦ Bob Nangle of Meridian printing ❦ Sandy, of Scraptastic—the world's leading specialist in the production of die-cut paper ripple chips ❦ The library staff at Brown University, and most especially William S. Monroe ❦ The librarians at the Providence Athenaeum, and in particular Christina Bevilacqua, Alison Maxell, and Lindsay Shaw ❦ Recycling for Rhode Island Education ❦ John Wemple ❦ Yo-Yo Mike at The YoStore. com in Makawao, HI ❦ And most especially to Liz Wells, tireless and exuberant Director of Meetings for the Snack Food Association **PERSONAL** Mark and Harry of Adler's Hardware ❦ The Allios ❦ Michael Boyer ❦ Aleksa Bzenic ❦ Sally Dowling ❦ The Dubno-Bernsteins ❦ Rick Engle ❦ Françoise Dussart ❦ Field Elementary School, Weston, MA ❦ Gordon School ❦ Sarah and Stephen Hall ❦ Olivia Hewitt ❦ Jolyon Howorth and Vivien Schmidt ❦ Hunnewell Elementary School, Wellesley, MA ❦ Edith Kurzweil ❦ The Long Island and Boston Kurzweils ❦ The Jordan Family ❦ The Joukowskys ❦ Jay Kernis ❦ Lincoln School ❦ Carrie Mauer, of the Providence Public School District ❦ Audrey, Claire, and Elliot Moylan ❦ Moses Brown School ❦ Paideia School, Atlanta, GA ❦ Ron Potvin and Chelsea Shriver ❦ Natasha and Kolya Markov-Riss ❦ The O'Shea-Pearlmans ❦ Ronald Schmidt ❦ Michael, Anne, and Amelia Spalter ❦ Alec, Liz, and Vivian Stansell ❦ Tenacre Country Day School, Wellesley, MA ❦ Quintin Viera ❦ And the whole Wheeler School community